The Erythrocyte - A Unique Cell

Edited by Vani Rajashekaraiah

Published in London, United Kingdom

The Erythrocyte - A Unique Cell
http://dx.doi.org/10.5772/intechopen.101007
Edited by Vani Rajashekaraiah

Contributors
Ahmed Shemran Mutlaq Alwataify, Ali Hadi Althabhawi, Hussain Naji Alshammary, Poonam Tripathi, Clemilson Berto Junior, Camila Cristina Guimarães-Nobre, Evelyn Mendonça-Reis, Lyzes Rosa Teixeira-Alves, Joan-Lluis Vives Corrons, Elena Krishnevskaya, Vani Rajashekaraiah, Masannagari Pallavi, Aastha Choudhary, Chaitra Bhat, Prerana Banerjee, Ranjithvishal, Shruthi Laavanyaa, Sudarshan Nithindran, Ashish Kumar Gupta, Shashi Bhushan Kumar, Erland Johansson, Anders B. Falk

First published in London, United Kingdom, 2023 by IntechOpen
IntechOpen is the global imprint of INTECHOPEN LIMITED, registered in England and Wales, registration number: 11086078, 5 Princes Gate Court, London, SW7 2QJ, United Kingdom

British Library Cataloguing-in-Publication Data
A catalogue record for this book is available from the British Library

Additional hard and PDF copies can be obtained from orders@intechopen.com

The Erythrocyte - A Unique Cell
Edited by Vani Rajashekaraiah
p. cm.
Print ISBN 978-1-80356-731-0
Online ISBN 978-1-80356-732-7
eBook (PDF) ISBN 978-1-80356-733-4

Meet the editor

Dr. Vani Rajashekaraiah is a professor in the Department of Biotechnology, Jain (Deemed to be University), India. She has 16 years of teaching experience in molecular biology, genetics, and genetic engineering, and 20 years of research experience in hematology and oxidative stress biology. She has a doctoral degree in Zoology from Bangalore University, India. Her current research focus is on blood storage solutions and drug-induced thrombocytopenia. Her research achievements include 5 funded research projects, 35 publications in reputed journals with 485 citations, 28 conference presentations, and one patent published in India in 2020. She received a Council of Scientific & Industrial Research (CSIR) fellowship in 2002, Seed Money to Young Scientists for Research from the Government of Karnataka in 2013, and the Jain University Achiever Award in 2015 and 2020 for quality research.

Contents

Preface

Erythrocytes are unique cells with characteristic structure and functions. As mature erythrocytes lack all the cell organelles, they have evolved distinct mechanisms of survival. This book examines some of the less-explored facets of erythrocytes in terms of their functions, diseases, and diagnostic markers.

The first section on erythrocyte functions introduces the characteristic attributes of erythrocytes as indicators of continually transforming conditions in their microenvironment. The second section on erythrocyte diseases gives insights into types of thalassemia and congenital defects due to membrane-related impairment.

I wish to thank Author Service Manager Maja Bozicevic and the rest of the team at IntechOpen for all their support throughout the publication process. I also acknowledge the authors for their excellent contributions as well as the chapter reviewers for their time and suggestions.

Vani Rajashekaraiah
Department of Biotechnology,
School of Sciences,
JAIN (Deemed-to-be University),
Bangalore, India

Section 1

Erythrocytes - Functions

Chapter 1

Reactive Oxygen Species and Antioxidant Interactions in Erythrocytes

Vani Rajashekaraiah, Masannagari Pallavi, Aastha Choudhary, Chaitra Bhat, Prerana Banerjee, Ranjithvishal, Shruthi Laavanyaa and Sudharshan Nithindran

Abstract

There is a continuous generation of reactive oxygen species (ROS) in erythrocytes due to their microenvironment. Reactive oxygen species (ROS) and reactive nitrogen species are well known as both harmful and beneficial species. They help in activating the antioxidant enzymes. However, overproduction of ROS can cause fatal damage to cell structures, including lipids and membranes, proteins and cause oxidative stress. Erythrocytes have effective antioxidant defenses to maintain their structure and functions. They protect these cells from damage and maintain their activities. Studies have reported that antioxidant interventions in various situations have proved beneficial to erythrocytes. Therefore, they can be employed as *in vitro* models for antioxidant and free radical interactions and also are ideal cell models for translational studies.

Keywords: erythrocytes, oxidative stress, free radicals, antioxidants, reactive oxygen species

1. Introduction

The erythrocyte [red blood cell (RBC)] is an ideal cell to study free-radical-mediated alterations. Approximately 25 trillion erythrocytes course through the human circulatory system. The main function of erythrocytes is the transport of oxygen (O_2) and the mediation of carbon dioxide (CO_2) production [1]. Reactive oxygen species (ROS) are continuously produced within the erythrocytes due to high O_2 tension in arterial blood and heme iron content [2].

The mature erythrocyte contains a variety of enzymes, proteins, carbohydrates, lipids, anions, and cations, to balance the cell's metabolism and functions. An important consequence of erythrocyte imbalance is a reduced ability to deal with oxidative stress, which can lead to degenerative changes in hemoglobin, membrane, and enzymes [3].

Erythrocytes are exposed to circulating inflammatory mediators and related oxidative stress, which cause severe alterations in cellular membrane and functions in a variety of pathological conditions. These alterations have been defined as "erythropathy" [4] and can be observed in conditions of cardiovascular injury [5, 6]. The loss of lipid asymmetry, and thus the exposure of phosphatidyl serine (PS) on the outer monolayer, contributes to the premature destruction of thalassaemic and sickle red cells [7, 8]. Sickle cell disease is distinguished by a change in erythrocyte shape from biconcave discs to elongated and sickle-shaped erythrocytes, consequently leading to loss of function and anemia [9].

Chronic obstructive pulmonary disease (COPD) causes changes in erythrocyte shape, redistribution of microfilaments such as actin and spectrin, and elevations in membrane rigidity [10]. Alterations were also observed in terms of erythrocyte morphology (leptocytes and elliptocytes), elevated membrane F2-isoprostanes and 4-hydroxynonenal (4-HNE) protein adducts, and oxidative damage to actin proteins in Rett syndrome (RTT) and autism spectrum disorder (ASD) [11, 12].

The changes in erythrocyte morphology and stiffness have also been reported in pathologies (type 2 diabetes, obesity, hypertension, and hypercholesterolemia) characterized by consistent oxidative damage followed by reshaping of the lipid distribution and architecture [13]. Erythrocytes participate in physiological and pathological processes associated with oxidative stress, such as aging, Down syndrome, neurodegenerative diseases such as Alzheimer's disease, erectile dysfunction, and cardiovascular disease [14].

The potential clinical application of these erythrocyte alterations as new biomarkers could be useful tools for monitoring a variety of oxidative-stress-related diseases.

2. Reactive oxygen species (ROS) in erythrocytes

Various physiological and pathological conditions, for example, aging, inflammation, and cell death develop through ROS generation. Several factors can lead to the generation of oxidizing radicals such as superoxide anion ($O_2^{\bullet-}$), hydrogen peroxide (H_2O_2), and hydroxyl radical ($HO^{\bullet}$) in erythrocytes [15].

Free radicals can be formed in three ways:

i. The cleavage of a covalent bond of a normal molecule, with each fragment retaining one of the paired electrons;

ii. The loss of a single electron from a normal molecule;

iii. The addition of a single electron to a normal molecule.

The latter, electron transfer, is a common process in biological systems [16].
Free radicals and ions are formed as illustrated below:
Radical formation by electron transfer: $A + e^- \rightarrow A^{\bullet-}$
Radical formation by homolytic fission: $X\!:\!Y \rightarrow X^{\bullet} + Y^{\bullet}$
Ion formation by heterolytic fission: $X\!:\!Y \text{ —> } X\!:^- + Y^+$

2.1 Nature of reactive oxygen species

ROS are defined as oxygen-containing species, which are highly reactive. O_2 undergoes one or two-electron reduction to form ROS, which reacts quickly with other compounds, attempting to capture the required electron in order to gain stability. ROS are oxygen-centered molecules that include hydrogen peroxide, singlet oxygen, superoxide anion, hydroxyl radical, and nitric oxide (NO) [16]. ROS are constantly produced in small quantities by normal metabolic processes. The addition of one electron to O_2 forms $O_2^{\bullet -}$, whereas the addition of two electrons results in the production of H_2O_2.

There are two causes for $O_2^{\bullet -}$ generation in erythrocytes.

Firstly, Oxyhemoglobin (oxyHb) autoxidizes at a relatively slow rate to yield methemoglobin (metHb), and $O_2^{\bullet -}$, which, further produces H_2O_2. Hemoglobin (Hb) is constantly exposed to an intracellular and extracellular flux of H_2O_2. When oxyHb is exposed to H_2O_2, it undergoes oxidative modifications that have been proposed as selective signals for proteolysis in erythrocytes [17]. Secondly, the oxidation state of trivalent iron (Fe^{3+}) has lost an electron during its formation; consequently, $O_2^{\bullet -}$ has been generated from exogenous sources, such as drugs, etc. [18].

Hydrogen peroxide is hydrophilic; however, recent studies reported that aquaporins are not involved in facilitating H_2O_2 diffusion across RBC membranes; rather, diffusion occurs through the lipid fraction or an unidentified membrane protein [19]. While charged, $O_2^{\bullet -}$ can only cross membranes via transmembrane anion channels. MetHb, lipid peroxidation, and spectrin-Hb complexes increase with H_2O_2, which further generates a covalent complex of spectrin and Hb, leading to changes in cell shape, membrane deformability, phospholipid organization, and cell surface characteristics [20].

The Fenton reaction occurs when H_2O_2 reacts with ferrous iron to produce $OH^{\bullet}$. H_2O_2 can react with $O_2^{\bullet -}$ to generate $OH^{\bullet}$, the most active ROS that cannot travel far due to its short half-life of a few nanoseconds known as Haber-Weiss Reaction [21, 22].

$$Fe^{++} + \mathrm{H_2O_2} \rightarrow Fe^{+++} + OH^{-} + OH^{\bullet} \,(\text{Fenton Reaction}) \quad (1)$$

$$\mathrm{O_2^{\bullet -}} + \mathrm{H_2O_2} \rightarrow \mathrm{O_2} + OH^{-} + OH^{\bullet} \,(\text{Haber-Weiss Reaction}) \quad (2)$$

ROS have the ability to act as both oxidizing and reducing agents. ROS are capable of directly attacking the red cell membrane and causing changes in lipid and protein structure [23]. ROS also alter mechanical properties, increase rigidity, and RBC interactions with other cells and coagulation factors, as well as stimulate microparticle (MP) generation and phosphatidylserine (PS) exposure [24]. Human red cell aging could be attributed to oxidative damage. RBC deformability, membrane permeability, and surface antigenicity abnormalities, on the other hand, have been recognized as defects in cellular properties that contribute to RBC senescence [25].

Nitric oxide, along with O_2 and CO_2, is the third gas transported by erythrocytes. Erythrocytes are the primary NO scavengers in circulation due to their high Hb concentration. NO is taken up by heme prosthetic groups of Hb-chain cysteine residues. NO is converted to nitrate by oxyhemoglobin ($HbFe^{+2}O_2$), whereas deoxyhemoglobin ($HbFe^{2+}$) binds to NO to form iron-nitrosylhemoglobin ($HbFe^{2+}NO$). NO consumption by erythrocytes can be regulated by $HbFe^{2+}NO$ formation under hypoxic conditions [26]. NO reaction with Hb greatly limits intravascular NO concentration. As a result, it is unlikely that NO is directly exported or produced by red blood cells as an intravascular signaling molecule. The rapid deoxygenation of NO by Hb results in the formation of nitrate and metHb, preventing NO diffusion from plasma to smooth muscle [15]. NO is produced in large amounts in inflammatory conditions and reacts with $O_2^{\bullet -}$ to generate peroxynitrite [27]. Peroxynitrite oxidizes plasma components, releasing secondary radicals that promote tyrosine nitration, leading to gain or loss of protein function [28].

3. Oxidative stress in erythrocytes

Erythrocytes are well endowed to combat oxidative stress due to their continuous contact with oxygen, as their inherent carrier function. When the concentration of ROS in cells or tissues exceeds the antioxidant protection, oxidative stress occurs [21]. RBC properties have been shown to change as a result of oxidative damage. Oxidative damage can also alter membrane permeability resulting in hemolysis [29, 30]. Oxidative cross-linking of spectrin can cause increased membrane rigidity and decreased erythrocyte deformability. Erythrocytes can be recognized by the immune system as a result of oxidative damage [31].

3.1 Extracellular hemoglobin as a source of oxidative stress

Extracellular Hb, which results from erythrocyte hemolysis or the infusion of cell-free Hb-based blood substitutes, can be a major source of oxidative stress. This potential source of oxidative stress is minimized under normal conditions by haptoglobin and hemopexin, which bind Hb and free heme, respectively. They inhibit the oxidative reactions of Hb and heme, allowing them to be removed from circulation. Elevated levels of free Hb and heme, which cannot be neutralized by haptoglobin and hemopexin, cause a variety of adverse clinical effects [32].

Autoxidation of Hb produces superoxide as well as methemoglobin. Hb is known to react with hydrogen peroxide to form ferrylhemoglobin, a strong oxidant [33]. Hb binds to RBC membrane proteins, especially under hypoxic conditions. The ROS produced by bound Hb may be inaccessible to cellular antioxidants, allowing the production of heme degradation products close to the membrane [34–36]. Extensive lipid peroxidation results in changes in fluidity such as a drop in membrane potential and an increase in permeability to different ions, which eventually leads to hemolysis. Thus, perturbations in erythrocyte function and structure can result in an increased flow of prooxidant generation that can lead to oxidative stress.

Oxidative stress and ROS accumulation in RBCs during aging may induce hemolysis. As a result, the plasma proteins haptoglobin and hemopexin can render free Hb and heme relatively inactive and deliver them safely to macrophages for phagocytosis [37, 38]. Oxidized Hb on the other hand exhibits impaired plasma clearance, due to its low affinity for haptoglobin protein.

Erythrocyte membranes exposed to oxidative stress undergo cellular component modifications such as oxidative denaturation of Hb, peroxidation of lipids, and high-molecular-weight cross-linked membrane proteins. The erythrocyte membrane is rich in sulfhydryl (−SH) groups, which help to maintain cellular oxidative balance [39]. Changes in the membrane elasticity may occur due to oxidative damage to the membrane −SH groups [40]. Oxidative damage to erythrocytes can occur and manifest itself in a variety of ways, including potassium release, increase in malondialdehyde (MDA), phosphatidylserine externalization [41], decrements in glutathione, superoxide dismutase, glutathione peroxidase, glutathione S-transferase, and glutathione reductase as well as total antioxidant activity of plasma [42–44]. Oxidant stress is a key component to both normal RBC aging and pathological dysfunction [32].

Oxidative damage to a specific protein, particularly at the active site, can result in the progressive loss of a specific biochemical function [45]. Peroxidation causes globin cross-linking to proteins such as spectrin and band 3. These processes further lead to decreased phospholipid symmetry, formation of cross-linked spectrin and Hb, aggregation of band 3 protein, and increase in advanced glycation end products leading to deformability and morphologic and surface changes in the erythrocyte [15, 46, 47]. During erythrocyte aging, an irreversible oxidative complex is formed between the Hb globin chain and spectrin [20]. Iron release may be accompanied by the generation of senescent antigens (SCA) and oxidative alteration of membrane proteins [15, 48, 49]. Another event associated with red cell homeostasis disruption is an increased inflammatory state in the bloodstream, since erythrocytes are constantly exposed to inflammatory molecules transported in the vascular system, their membranes may be particularly vulnerable to their interaction. A large number of inflammatory mediators, including tumor necrosis factor-α, interleukins, interferons, and C-reactive protein (CRP), have been proposed as potential inflammatory response markers [50].

In general, erythrocytes respond to oxidative stress by activating tyrosine kinases [51, 52], resulting in tyrosine phosphorylation at the cytoplasmic domain of band 3 protein, which mediates interactions with ankyrin, leading to membrane destabilization. This also seen in β-thalassemia disease with the loss of stability between cytoskeleton and membrane complexes, such as band 3 protein [53]. Oxidative damage to band 3 has been linked to RBC aging including the exposure of senescent specific neo-antigens that bind autologous IgG triggering RBC removal [54].

Erythrocytes also contain NADH oxidases, which can generate endogenous ROS [55]. However, some forms of NADH oxidase were also detected in normal RBCs. L-isoaspartyl groups were discovered on erythrocyte membrane proteins in response to aging or pathological oxidative stress, such as glucose 6-phosphate dehydrogenase deficiency or Down syndrome [56–59]. Thus, L-isoaspartyl group accumulation in RBC proteins correlates with RBC dysfunction and pathology.

The effects of oxidative stress observed in various pathological conditions in erythrocytes are depicted in **Table 1**.

Diseases	Oxidative stress markers	Reference
Sickle cell disease	Variations in Superoxide dismutase	[60–63]
	Alterations in Catalase	[60, 62, 63]
	Glutathione—decreased	[64, 65]
	Changes in glutathione peroxidase	[61, 62]
	Vitamin E—decreased	[66, 67]
	Vitamin-C—decreased	[68]
	Lipid peroxidation—increased	[62, 63, 69, 70]
	Intracellular Ca^{2+}—increased	[71]
	Phosphatidylserine exposure—increased	[72–74]
	NADPH oxidase activity—increased	[55]
	ROS generation—increased	[55]

Diseases	Oxidative stress markers	Reference
β-Thalassemia	Band 3 tyrosine phosphorylation (P)—increased	[53]
	Phosphatidylserine exposure—increased	[53]
	Release of microparticles	[53]
	Membrane thiols—decreased	[53]
	K^+, Cl^-, and water loss	[53]
Diabetes (T2DM)	Lipid peroxidation—increased	[75]
	GSH/GSSG ratio—decreased	[75]
	Glutathione—decreased	[75]
	Phosphatidylserine externalization—increased	[75]
	Annexin binding—increased	[75]
	Caspase-3 activation—increased	[76]
Malaria	Lipid peroxidation—increased	[77]
	Glutathione—decreased	[77]
	Catalase—decreased	[77]
	Membrane stiffness—increased	[78]
Autism	TBARS—increased	[79]
	Xanthine oxidase—increased	[79]
	Superoxide dismutase—increased	[79]
	Superoxide dismutase—decreased	[80]
	Glutathione peroxidase—decreased	[80]
	Modulations in catalase	[79, 81]
Chronic kidney disease	Membrane fluidity—increased	[82]
	Osmatic fragility—increased	[82]
	RBC life span—decreased	[82]
	Antioxidant enzymes—decreased	[83–85]
	NADPH oxidase—increased	[86]
	Hemoglobin leakage—increased	[86]
	Nitrite ions—increased	[86]
	Peroxynitrite—increased	[86]
	Hemoglobin—decreased	[86]
	Circulating RBCs—decreased	[86]
Systemic sclerosis (SSc)	Anion exchange capability via band 3—decreased	[87]
	Blood viscosity—increased	[87]
	Membrane protein structure destabilization	[87]
	SO_4^{2-}—decreased	[87]

Diseases	Oxidative stress markers	Reference
Corona virus disease-19 (Covid-19)	Hematocrit levels—decreased	[88]
	RBCs amplitude—increased	[88]
	Glutathione—decreased	[89–91]
	Arginase 1—increased	[92]
	ROS—increased	[92]
	Interferon-γ—increased	[92]
	NO bioactivity—decreased	[92]
	Oxidized glutathione (GSSG)—increased	[93]
	Glycolytic metabolites—increased	[93]
	Phosphoglucomutase—decreased	[93]
	Carboxylic acids—increased	[93]
	Glutathione peroxidase—decreased	[91]
	Catalase—decreased	[91]
	Superoxide dismutase—decreased	[91]

Table 1.
Effects of oxidative stress in erythrocytes during diseases.

4. Antioxidant defense in erythrocytes

The antioxidant defense system of erythrocytes prevents oxidative cell damage. This implies that the erythrocyte antioxidant defenses operate in a balanced manner. As a result, an appropriate redox state, balanced antioxidant action is required for ROS homeostasis.

Antioxidants are molecules that prevent or delay cellular damage by inhibiting or quenching ROS reactions [94]. Antioxidants can be synthesized in the body or obtained from the environment, such as through diet. Erythrocytes are well equipped to fight against oxidative stress, i.e., mechanisms to scavenge and detoxify ROS, prevent their production, and sequester transition metals [95]. Erythrocytes contain both enzymatic and non-enzymatic antioxidants to combat oxidative stress.

4.1 Enzymatic antioxidants

Enzymatic antioxidants include superoxide dismutase (SOD), catalase (CAT), glutathione peroxidase (GPX), glutathione reductase (GR), and peroxiredoxin-2 (PRX-2). Their coordinated actions protect the erythrocytes from free-radical-mediated damage. Since there is no *de novo* synthesis of antioxidant enzymes in mature erythrocytes, their defense capacity is limited. Free radicals affect the capacities of antioxidative enzymes as well as the overall antioxidative system [96]. Under normal conditions, erythrocytes contain sufficient levels of scavenger enzymes such as Cu,Zn-SOD, CAT, and selenium-dependent GPX to protect from free radical injury.

Superoxide dismutase, a ubiquitous metal-containing enzyme, involves in the detoxification of $O_2^{\bullet-}$ into O_2 and H_2O_2.

$$O_2^{\bullet -} + O_2^{\bullet -} + 2H^+ \xrightarrow{SOD} H_2O_2 + O_2 \quad (3)$$

SOD family comprises CuSOD, ZnSOD, MnSOD, and extracellular SOD, which protect from particularly $O_2^{\bullet -}$. Cu,Zn-SOD catalyzes the dismutation of $O_2^{\bullet -}$ to H_2O_2, which is later converted to water by CAT or GPX [21, 27, 97]. The activity of these enzymes in erythrocytes is highest than that of other tissues in the body [98].

Erythrocytes are well protected against ROS due to the abundance of Cu,Zn-SOD, which scavenges free radicals and thus prevents metHb formation [43]. Cu, Zn-SOD synthesis is induced by $O_2^{\bullet -}$ formation through the activation of regulatory genes [15]. SOD scavenges $O_2^{\bullet -}$ and inhibits the formation of peroxynitrite, thereby preventing injury and regulating the bioavailability of NO [27]. Erythrocytes contain an abundant quantity of Cu,Zn-SOD, which maintains intra-erythrocyte $O_2^{\bullet -}$ levels at concentrations as low as 10^{-13} mol/L [96].

Catalases catalyze the direct decomposition of H_2O_2 to water and O_2 [21].

$$2H_2O_2 \xrightarrow{CAT} 2H_2O + O_2 \quad (4)$$

Catalase and SOD react synergistically to protect each other [99]. CAT and GPX are equally active in the detoxification of H_2O_2 in normal erythrocytes [100]. At physiological concentrations, GPX acts as a primary defense in H_2O_2 degradation by reducing H_2O_2 while also converting GSH to its oxidized form (GSSG). However, under H_2O_2 overproduction, CAT exhibits increased enzymatic activity, as measured by the Michaelis-Menten constant (Km). The Km for CAT (2.4×10^{-4} M) is significantly greater than the Km for GPX (1×10^{-6} M), which indicates that CAT scavenges H_2O_2 efficiently at higher concentrations [101–103]. GPX is important in dealing with endogenous H_2O_2 produced by Hb autoxidation, whereas CAT becomes increasingly important when erythrocytes are exposed to increased H_2O_2 flux [15]. H_2O_2 readily crosses erythrocyte membranes and can protect other tissues against extracellular H_2O_2 by "absorbing" and destroying it. Reduced glutathione (GSH) is used by GPX to detoxify hydrogen peroxide during normal antioxidant defense system function. Furthermore, glutathione reductase is required to convert H_2O_2 to GSH, which contributes to H_2O_2 detoxification [104].

Glutathione-S-transferases are important in the detoxification of electrophilic xenobiotics. This enzyme catalyzes the conjugation of GSH with exogenous and endogenous toxic compounds or their metabolites, making them more water-soluble, less toxic, and easier to excrete. In addition, they are responsible for various resistance mechanisms such as chemotherapeutic or antibiotic drug resistance [15].

Glucose-6-phosphate dehydrogenase (G6PD) is an important antioxidant enzyme in erythrocytes, which is the regulatory enzyme of the pentose-phosphate pathway (PPP). As erythrocytes lack mitochondria, the PPP pathway is the only source of NADPH, and it plays an important role in NADPH-dependent antioxidant defense [105, 106]. G6PD is required to protect erythrocytes from oxidative damage. The lack of this protection can lead to severe hemolysis [107].

In addition to primary antioxidant defense systems that prevent the generation of free radicals or radical chain reactions, secondary systems have been proposed. Proteases that preferentially degrade oxidatively damaged proteins are among them. In erythrocytes, a multicatalytic proteolytic complex appears to be responsible for the degradation of oxidized intracellular proteins [47]. The presence of an 80-kDa serine protease in the oxidized erythrocyte membranes preferentially degrades oxidized proteins specifically protein hydrolase. When cells are oxidized, this cytoplasmic protein becomes adherent to membranes, promoting membrane protein degradation. The protease is characterized by its inhibition by a serine protease inhibitor [108]. It is endogenously present in oxidized or aged erythrocyte membranes and plays a crucial role in the removal of the oxidation-induced membrane protein aggregates and in reducing the oxidation-induced anti-band 3 binding in aging. Oxidized protein hydrolase (OPH) acts as a secondary defense system by removing oxidized protein aggregates.

Peroxiredoxins (PRX), a class of thiol-containing enzymes, act as H_2O_2 and peroxynitrite scavengers in circulation. PRXs have a reductive capacity for hydroperoxides via a reductant thiol. Peroxiredoxins have been shown in studies to be catalytic peroxynitrite reductases. It has been also reported that PRX-II is present in the cytosol of erythrocytes. The catalytic cycle involves the reduction of oxidized PRX by thioredoxin and the reduction capacity of NADPH via NADPH-thioredoxin reductase [28, 109].

4.2 Non-enzymatic antioxidants

Endogenous non-enzymatic antioxidants are defined in two phases: lipophylic (vitamin E, carotenoids, ubiquinon, melatonin, etc.) and water-soluble (vitamin C, glutathione, uric acid, ceruloplasmin, transferin, haptoglobulin, etc.). Three antioxidant vitamins, A, C, and E, provide defense against oxidative damage. Vitamin C acts in the aqueous phase, whereas vitamin E acts in the lipid phase as a chain-breaking antioxidant. Vitamin C reduces $O_2^{\bullet -}$ and lipid peroxyl radical, but is also a well-known synergistic agent for vitamin E [2]. Uric acid is an endogenous antioxidant with metal-chelating properties and scavenges nitrogen radicals and superoxide in plasma, thereby blocking the generation of peroxynitrite. Uric acid in erythrocytes quenches the free radicals and ROS. Uric acid maintains the smooth membrane surface of RBCs, thus preventing echinocyte formation [110].

Additionally, erythrocytes have a plasma membrane redox system (PMRS) that transfers electrons from intracellular substrates to extracellular electron acceptors, which may be NAD+ or/and vitamin C [111].

Many studies have demonstrated the influence of different antioxidants on erythrocytes during oxidative stress (**Table 2**).

Sample	OS inducer	Antioxidant	Results	Reference
Human RBC membrane	Hydrogen peroxide (H_2O_2) & ozone	1,3-dimethyluric acid and 1,3,7-trimethyluric acid	Prevented lipid peroxidation	[112]
Human RBCs	Hydrogen peroxide	β-Carotene & resveratrol	Increased SOD & catalase. decreased conjugate dienes & TBARS	[113]
Human RBCs	Tert-butyl hydroperoxide	Resveratrol	Increase in GSH & membrane –SH	[114]
Human RBCs	tert-butyl hydroperoxide	Capsaicin or L-ascorbic acid	Decrease in MDA & protein carbonyls	[115]
Human RBCs	Hydrogen peroxide	Tea polyphenols (TPP)	Protected RBCs and membranes against lipid peroxidation	[116]
Sickle cell human RBCs	Tert-butyl hydroperoxide	Flavonoids (quercetin & rutin), Ascorbic acid	Protected against oxidative stress in and lipid peroxidation	[117]

Sample	OS inducer	Antioxidant	Results	Reference
Human RBCs	Hydrogen peroxide	Oak barrel-aged red wine extract (SD95)	Protected against hemolysis, ROS & maintained MDA	[118]
Human RBCs	AAPH or H_2O_2 or t-BOOH	Ether and water fraction of honey	Ether fraction inhibited hemolysis and t-BOOH-induced lipid peroxidation	[119]
Human RBCs	Bisphenol A (BPA)	Quercetin	Decreased MDA levels and increased antioxidant enzymes	[120]
Human RBCs	Hydrogen peroxide	Melatonin	Restored band 3 expression levels, cell shape alterations & lower TBARS.	[121]
Rat RBCs	Tert-butyl hydroperoxide	Flavonoids (quercetin, catechin and naringenin)	Inhibited membrane lipid peroxidation and decreased glutathione oxidation	[122]
Human RBCs	Sodium fluoride (NaF)	3,4-dihydroxy benzaldehyde (DHB)	Increased AOPP, lipid peroxidation, restored PMRS & AO enzyme activity	[123]
Human RBCs	Hydrogen peroxide	β-carotene (BC) and resveratrol (RSV)	TBARS and conjugate dienes decreased in BC and RSV groups. SOD increased in RSV.	[113]

Table 2.
The effects of antioxidants on erythrocytes during oxidative stress.

There are *in vitro* studies on oxidative stress in erythrocytes reporting the protective effects of antioxidants from plant extracts (**Table 3**).

Sample	OS inducer	Antioxidant	Results	Reference
Rat RBCs	H_2O_2	Potato peel extract (PPE)	Inhibited morphological alterations RBCs	[124]
Human RBCs	2,2′-azobis (2-amidinopropane) dihydrochloride (AAPH)	Olive leaf extract (OLE)	Inhibited hemolysis, TBARS formation, and hemoglobin oxidation	[125]
Goat RBCs	Cu^{2+}-ascorbate	Terminalia arjuna (TA) bark extract	Decreased lipid peroxidation and protein carbonyl content. Increased GSH	[126]
Human RBCs	2,2′-azobis (2-amidinopropane) dihydrochloride (AAPH)	Quince (Cydonia oblonga) fruit pulp and peel extracts	Protection of erythrocyte membrane from hemolysis	[127]
Human RBCs	Hypochlorous acid (HClO)	Ugni molinae Turcz aqueous extract	Reduced hemolysis	[128]
Rat RBCs	Cadmium	Salicornia arabica lipid extract	Ameliorated antioxidant status and inhibited MDA levels	[129]

Sample	OS inducer	Antioxidant	Results	Reference
Human RBCs	Hypochlorous acid (HClO)	Pitavia punctata extract	Protecting the membrane integrity and inhibiting the oxidation of the LDL lipoprotein	[130]
Human RBCs	H_2O_2	Orchis latifolia and Centratherum anthelminticum extract	Protected membrane integrity resulting in a reduction of RBC hemolysis and lipid peroxidation	[131]

Table 3.
The modulations of antioxidants from plant extracts in erythrocytes.

5. Conclusion

Many physiological and pathological circumstances have the potential to cause oxidative stress and are possibly harmful in susceptible individuals. Furthermore, this increased risk of adverse reactions is generally reflected in the erythrocytes. The administration of antioxidants has the prospects of diminishing oxidative damage. Therefore, erythrocytes act as unique cell models for translational studies on oxidant and antioxidant interactions. However, they may not be helpful to study these effects in relation to mitochondria (a major source of ROS), as the mature erythrocytes lack the cell organelles.

Acknowledgements

The authors acknowledge Dr. Leela Iyengar, Anusha BA, Ms. Magdaline Christina, and JAIN (Deemed-to-be University) for their support.

Conflict of interest

The authors have no conflict of interest to disclose.

Author details

Vani Rajashekaraiah*, Masannagari Pallavi, Aastha Choudhary, Chaitra Bhat, Prerana Banerjee, Ranjithvishal, Shruthi Laavanyaa and Sudharshan Nithindran
Department of Biotechnology, School of Sciences, JAIN (Deemed-to-be University), Bangalore, India

*Address all correspondence to: vani.rs@jainuniversity.ac.in; tiwari.vani@gmail.com

References

[1] Volpe EP. Blood and circulation. In: Dubuque WC, editor. Biology and Human Concerns. Dubuque: Wm.C.Brown Publishers; 1993. pp. 253-265

[2] Baynes JW. Oxygen and life. In: Baynes JW, Domoniczak MH, editors. Medical Biochemistry. Philadelphia: Elsevier; 2005. pp. 497-506

[3] Edwards JC, Fuller J. Oxidative stress in erythrocytes. Comparative Haematology International. 1996;**6**: 24-31. DOI: 10.1007/BF00368098

[4] Pernow J, Mahdi A, Yang J, Zhou Z. Red blood cell dysfunction: A new player in cardiovascular disease. Cardiovascular Research. 2019;**115**:1596-1605. DOI: 10.1093/cvr/cvz156

[5] Zhou Z, Mahdi A, Tratsiakovich Y, Zahorán S, Kövamees O, Nordin F, et al. Erythrocytes from patients with type 2 diabetes induce endothelial dysfunction via arginase I. Journal of the American College of Cardiology. 2018;**72**:769-780. DOI: 10.1016/j.jacc.2018.05.052

[6] Yang J, Zheng X, Mahdi A, Zhou Z, Tratsiakovich Y, Jiao T, et al. Red blood cells in type 2 diabetes impair cardiac post-ischemic recovery through an arginase-dependent modulation of nitric oxide synthase and reactive oxygen species. JACC: Basic to translational. Science. 2018;**3**:450-463. DOI: 10.1016/j.jacbts.2018.03.006

[7] Kuypers FA, Yuan J, Lewis RA, Snyder LM, Kiefer CR, Bunyaratvej A, et al. Membrane phospholipid asymmetry in human thalassemia. Blood. The Journal of the American Society of Hematology. 1998;**91**:3044-3051. DOI: 10.1182/blood.v91.8.3044.3044_3044_3051

[8] Setty BY, Kulkarni S, Stuart MJ. Role of erythrocyte phosphatidylserine in sickle red cell–endothelial adhesion. Blood. The Journal of the American Society of Hematology. 2002;**99**:1564-1571. DOI: 10.1182/blood.V99.5.1564

[9] Kato GJ, Piel FB, Reid CD, Gaston MH, Ohene-Frempong K, Krishnamurti L, et al. Sickle cell disease. Nature Reviews Disease Primers. 2018;**4**:1-22. DOI: 10.1038/nrdp.2018.10

[10] Santini MT, Straface E, Cipri A, Peverini M, Santulli M, Malorni W. Structural alterations in erythrocytes from patients with chronic obstructive pulmonary disease. Pathophysiology of Haemostasis and Thrombosis. 1997;**27**:201-210. DOI: 10.1159/000217458

[11] Ciccoli L, De Felice C, Paccagnini E, Leoncini S, Pecorelli A, Signorini C, et al. Morphological changes and oxidative damage in Rett syndrome erythrocytes. Biochimica et Biophysica Acta (BBA)-General Subjects. 2012;**1820**:511-520. DOI: 10.1016/j.bbagen.2011.12.002

[12] Ciccoli L, De Felice C, Paccagnini E, Leoncini S, Pecorelli A, Signorini C, et al. Erythrocyte shape abnormalities, membrane oxidative damage, and β-actin alterations: An unrecognized triad in classical autism. Mediators of Inflammation. 2013;**2013**. Article ID 432616. DOI: 10.1155/2013/432616

[13] Gyawali P, Richards RS, Uba NE. Erythrocyte morphology in metabolic syndrome. Expert Review of Hematology. 2012;**5**:523-531. DOI: 10.1586/ehm.12.47

[14] Massaccesi L, Galliera E, Corsi Romanelli MM. Erythrocytes as markers of oxidative stress-related pathologies. Mechanisms of Ageing

and Development. 2020;**191**:111333. DOI: 10.1016/j.mad.2020.111333

[15] Cimen MYB. Free radical metabolism in human erythrocytes. Clinica Chimica Acta. 2008;**390**:1-11. DOI: 10.1016/j.cca.2007.12.025

[16] Cheeseman KH, Slater TF. An introduction to free radical biochemistry. British Medical Bulletin. 1993;**49**:481-493. DOI: 10.1093/oxfordjournals.bmb.a072625

[17] Giulivi C, Davies KJ. Mechanism of the formation and proteolytic release of H2O2-induced dityrosine and tyrosine oxidation products in haemoglobin and red blood cells. The Journal of Biological Chemistry. 2001;**276**:24129-24136. DOI: 10.1074/jbc.M010697200

[18] Hebbel RP, Eaton JW, Balasingam M, Steinberg MH. Spontaneous oxygen radical generation by sickle erythrocytes. The Journal of Clinical Investigation. 1982;**70**:1253-1259. DOI: 10.1172/jci110724

[19] Orrico F, Lopez AC, Saliwonczyk D, Acosta C, Rodriguez-Grecco I, Mouro-Chanteloup I, et al. The permeability of human red blood cell membranes to hydrogen peroxide is independent of aquaporins. Journal of Biological Chemistry. 2022;**298**:101503. DOI: 10.1016/j.jbc.2021.101503

[20] Snyder LM, Fortier NL, Trainor J, Jacobs J, Leb L, Lubin B, et al. Effect of hydrogen peroxide exposure on normal human erythrocyte deformability, morphology, surface characteristics, and spectrin-hemoglobin cross-linking. Journal of Clinical Investigation. 1985;**76**:1971-1977. DOI: 10.1172/JCI112196

[21] Al-Omar MA, Beedham C, Alsarra IA, Mohammed AAO. Pathological roles of reactive oxygen species and their defence mechanisms. Saudi Pharmaceutical Journal. 2004;**12**:1-18. ID: emr-69011

[22] Yan EB, Unthank JK, Castillo-Melendez M, Miller SL, Langford SJ, Walker DW. Novel method for in vivo hydroxyl radical measurement by microdialysis in fetal sheep brain in utero. Journal of Applied Physiology. 2005;**98**:2304-2310. DOI: 10.1152/japplphysiol.00617.2004

[23] Claster S, Chiu DT, Quintanilha A, Lubin B. Neutrophils mediate lipid peroxidation in human red cells. Blood. 1984;**64**:1079-1084

[24] Wang Q, Zennadi R. Oxidative stress and thrombosis during aging: The roles of oxidative stress in RBCs in venous thrombosis. International Journal of Molecular Sciences. 2020;**21**:4259. DOI: 10.339 0/ijms21124259

[25] Chiu D, Lubin B. Oxidative hemoglobin denaturation and RBC destruction: The effect of heme on red cell membranes. Seminars in Hematology. 1989;**126**:128-135

[26] Han TH, Qamirani E, Nelson AG, Hyduke DR, Chaudhuri G, Kuo L, et al. Regulation of nitric oxide consumption by hypoxic red blood cells. Proceedings of the National Academy of Sciences. 2003;**100**:12504-12509. DOI: 10.1073/pnas.2133409100

[27] Gunduz K, OzturkG SEY. Erythrocyte superoxide dismutase, catalase activities and plasma nitrite and nitrate levels in patients with Behçet disease and recurrent aphthous stomatitis. Clinical and Experimental Dermatology. 2004;**29**:176-179. DOI: 10.1111/j.1365-2230.2004.01488.x

[28] Romero N, Denicola A, Radi R. Red blood cells in the metabolism of

nitric oxide-derived peroxynitrite. IUBMB Life. 2006;**58**:572-580. DOI: 10.1080/15216540600936549

[29] Lubin B, Chiu DTY. Properties of vitamin E deficient erythroeytes following peroxidant injury. Pediatric Research. 1982;**16**:928-932. DOI: 10.1203/00006450-198211000-00005

[30] Durak I, Kavutcu M, Çimen MYB, Avc A, Elgün S, Öztürk HS. Oxidant/antioxidant status of erythrocytes from patients with chronic renal failure: Effects of hemodialysis. Medical Principles and Practice. 2001;**10**:187-190. DOI: 10.1159/000050367

[31] Low P, Waugh SM, Zinke K, Drenckhahan D. The role of hemoglobin denaturation and band 3 clustering in red blood cell aging. Science. 1985;**227**:531-533. DOI: 10.1126/science.2578228

[32] Rifkind JM, Mohanty JG, Nagababu E. The pathophysiology of extracellular hemoglobin associated with enhanced oxidative reactions. Frontiers in Physiology. 2015;**5**:500. DOI: 10.3389/fphys.2014.00500

[33] Everse J. Hsia N the toxicities of native and modified hemoglobins. Free Radical Biology and Medicine. 1997;**22**:1075-1099. DOI: 10.1016/s0891-5849(96)00499-6

[34] Bakonyi T, Radak Z. High altitude and free radicals. Journal of Sports Science and Medicine. 2004;**3**:64-69

[35] Nagababu E, Rifkind JM. Heme degradation during autoxidation of oxyhemoglobin. Biochemical and Biophysical Research Communications. 2000;**273**:839-845. DOI: 10.1006/bbrc.2000.3025

[36] Nagababu E, Chrest F, Rifkind J. Hydrogen-peroxide-induced heme degradation in red blood cells: The protective roles of catalase and glutathione peroxidase. Biochimica et Biophysica Acta (BBA)—General Subjects. 2003;**1620**:211-217. DOI: 10.1016/s0304-4165(02)00537-8

[37] Smith A, Morgan WT. Haem transport to the liver by haemopexin. Receptor-mediated uptake with recycling of the protein. Biochemical Journal. 1979;**182**:47-54. DOI: 10.1042/bj1820047

[38] Thomsen JH, Etzerodt A, Svendsen P, Moestrup SK. The haptoglobin-CD163-heme oxygenase-1 pathway for hemoglobin scavenging. Oxidative Medicine and Cell Longevity. 2013;**2013**:Article ID:523652. DOI: 10.1155/2013/523652

[39] Reglinski J, Hoey S, Smith WE, Sturrock RD. Cellular response to oxidative stress at sulfhydryl group receptor sites on the erythrocyte membrane. Journal of Biological Chemistry. 1988;**263**:12360-12366. DOI: 10.1016/S0021-9258(18)37763-9

[40] Wang X, Wu Z, Song G, Wang H, Long M, Cai S. Effects of oxidative damage of membrane protein thiol groups on erythrocyte membrane viscoelasticities. Clinical Hemorheology and Microcirculation. 1999;**21**:137-146

[41] Chin-Yee I, Arya N, d'Almeida MS. The red cell storage lesion and its implication for transfusion. Transfusion Science. 1997;**18**:447-458. DOI: 10.1016/S0955-3886(97)00043-X

[42] Racek J, Herynková R, Holeček V, Jerabek Z, Slama V. Influence of antioxidants on the quality of stored blood. Vox Sanguinis. 1997;**72**:16-19. DOI: 10.1046/j.1423-0410.1997.00016.x

[43] Dumaswala UJ, Zhuo L, Jacobsen DW, Jain SK, Sukalski KA.

Protein and lipid oxidation of banked human erythrocytes: Role of glutathione. Free Radical Biology and Medicine. 1999;**27**:1041-1049. DOI: 10.1016/s0891-5849(99)00149-5

[44] Knight JA, Voorhees RP, Martin L, Anstall H. Lipid peroxidation in stored red cells. Transfusion. 1992;**32**:354-357. DOI: doi.org/10.1046/j.1537-2995.1992.32492263451.x

[45] Dubinina EE, Gavrovskaya SV, Kuzmich EV, Leonova NV, Morozova MG, Kovrugina SV, et al. Oxidative modification of proteins: Oxidation of tryptophan and production of dityrosine in purified proteins using fenton's system. Biochemistry. 2002;**67**:343-350. DOI: 10.1023/a:1014840617890

[46] Caprari P, Scuteri A, Salvati AM, Bauco C, Cantafora A, Masella R, et al. Aging and red blood cell membrane: A study of centenarians. Experimental Gerontology. 1999;**34**:47-57. DOI: 10.1016/s0531-5565(98)00055-2

[47] Fujino T, Tada T, Hosaka T, Beppu M, Kikugawa K. Presence of oxidized protein hydrolase in human cell lines, rat tissues, and human/rat plasma. The Journal of Biochemistry. 2000;**127**:307-313. DOI: 10.1093/oxfordjournals.jbchem.a022608

[48] Morabito R, Romano O, La Spada G, Marino A. H2O2-induced oxidative stress affects SO4= transport in human erythrocytes. PLoS One. 2016;**11**:e0146485. DOI: 10.1371/journal.pone.0146485

[49] Van Zwieten R, Verhoeven AJ, Roos D. Inborn defects in the antioxidant systems of human red blood cells. Free Radical Biology and Medicine. 2014;**67**:377-386. DOI: 10.1016/j.freeradbiomed.2013.11.022

[50] Remigante A, Morabito R, Marino A. Band 3 protein function and oxidative stress in erythrocytes. Journal of Cellular Physiology. 2021;**236**:6225-6234. DOI: 10.1002/jcp.30322

[51] Minetti M, Mallozzi C, Di Stasi AM. Peroxynitrite activates kinases of the src family and upregulates tyrosine phosphorylation signaling. Free Radical Biology and Medicine. 2002;**33**:744-754. DOI: 10.1016/s0891-5849(02)00891-2

[52] Mladenov M, Gokik M, Hadzi-Petrushev N, Gjorgoski I, Jankulovski N. The relationship between antioxidant enzymes and lipid peroxidation in senescent rat erythrocytes. Physiological Research. 2015;**64**:891-896. DOI: 10.33549/physiolres

[53] De Franceschi L, Bertoldi M, Matte A, Santos Franco S, Pantaleo A, Ferru E, et al. Oxidative stress and beta-thalassemic erythroid cells behind the molecular defect. Oxidative Medicine and Cellular Longevity. 2013;**2013**:Article ID: 985210. DOI: 10.1155/2013/985210

[54] Kay MM. Generation of senescent cell antigen on old cells initiates IgG binding to a neo antigen. Cellular and Molecular Biology. 1993;**39**:131-153

[55] George A, Pushkaran KDG, Koochaki S, Malik P, Mohandas N, Zheng Y, et al. Erythrocyte NADPH oxidase activity modulated by Rac GTPases, PKC,and plasma cytokines contributes to oxidative stress in sickle cell disease. Blood. 2013;**121**:2099-2107. DOI: 10.1182/blood-2012-07-441188

[56] Galletti P, Manna C, Ingrosso D, Iardino P, Zappia V. Hypotheses on the physiological role of enzymatic protein methyl esterification using human erythrocytes as a model system. Advances in Experimental Medicine

and Biology. 1991;**307**:149-160. DOI: 10.1007/978-1-4684-5985-2_14

[57] Galletti P, De Bonis ML, Sorrentino A, Raimo M, D'Angelo S, Scala I, et al. Accumulation of altered aspartyl residues in erythrocyte proteins from patients with Down's syndrome. The FEBS Journal. 2007;**274**:5263-5277. DOI: 10.1111/j.1742-4658.2007.06048.x

[58] Ingrosso D, Cimmino A, D'Angelo S, Alfinito F, Zappia V, Galletti P. Protein methylation as a marker of aspartate damage in glucose-6-phosphate dehydrogenase-deficient erythrocytes: Role of oxidative stress. European Journal of Biochemistry. 2002;**269**:2032-2039. DOI: 10.1046/j.1432-1033.2002.02838.x

[59] Janson CA, Clarke S. Identification of aspartic acid as a site of methylation in human erythrocyte membrane proteins. The Journal of Biological Chemistry. 1980;**255**:11640-11643. DOI: doi.org/10.1016/S0021-9258(19)70177-X

[60] Das SK, Nair RC. Superoxide dismutase, glutathione peroxidase, catalase and lipid peroxidation of normal and sickled erythrocytes. British Journal of Haematology. 1980;**44**:87-92. DOI: 10.1111/j.1365-2141.1980.tb01186.x

[61] Gizi A, Papassotiriou I, Apostolakou F, Lazaropoulou C, Papastamataki M, Kanavaki I, et al. Assessment of oxidative stress in patients with sickle cell disease: The glutathione system and the oxidant antioxidant status. Blood Cells Molecules and Diseases. 2011;**46**:220-225. DOI: 10.1016/j.bcmd.2011.01.002

[62] Alsultan AI, Seif MA, Amin TT, Naboli M, Alsulima AM. Relationship between oxidative stress, ferritin and insulin resistance in sickle cell disease. European Review for Medical and Pharmacological Sciences. 2010;**14**:527-538

[63] Dasgupta T, Hebbel RP, Kaul DK. Protective effect of arginine on oxidative stress in transgenic sickle mouse models. Free Radical Biology and Medicine. 2006;**41**:1771-1780. DOI: 10.1016/j.freeradbiomed.2006.08.025

[64] Reid M, Badaloo A, Forrester T, Jahoor F. In vivo rates of erythrocyte glutathione synthesis in adults with sickle cell disease. Endocrinology and Metabolism: American Journal of Physiology. 2006;**291**:E73-E79. DOI: 10.1152/ajpendo.00287.2005

[65] Morris C, Suh J, Hagar W, Larkin S, Bland D, Steinberg M, et al. Erythrocyte glutamine depletion, altered redox environment, and pulmonary hypertension in sickle cell disease. Blood. 2008;**111**:402-410. DOI: 10.1182/blood-2007-04-081703

[66] Tatum VL, Chow CK. Antioxidant status and susceptibility of sickle erythrocytes to oxidative and osmotic stress. Free Radical Biology and Medicine. 1996;**25**:133-139. DOI: 10.3109/10715769609149918

[67] Chiu D, Lubin B. Abnormal vitamin E and glutathione peroxidase levels in sickle cell anemia: Evidence for increased susceptibility to lipid peroxidation in vivo. Journal of Laboratory and Clinical Medicine. 1979;**94**:542-548

[68] Jain SK, Williams DM. Reduced levels of plasma ascorbic acid (vitamin C) in sickle cell disease patients: Its possible role in the oxidant damage to sickle cells in vivo. Clinica Chimica Acta. 1985;**149**:257-261. DOI: 10.1016/0009-8981(85)90339-0

[69] Silva DG, Belini JE, Torres LS, Ricci JO, Lobo CC, Bonini-Domingos CR, et al. Relationship between oxidative stress, glutathione S-transferase polymorphisms and hydroxyurea

treatment in sickle cell anemia. Blood Cells Molecules and Diseases. 2011;**47**:23-28

[70] Klings ES, Farber HW. Role of free radicals in the pathogenesis of acute chest syndrome in sickle cell disease. Respiratory Research. 2001;**2**:280-285

[71] Jain SK, Shohet SB. Calcium potentiates the peroxidation of erythrocyte membrane lipids. Biochimica et Biophysica Acta. 1981;**642**:46-54. DOI: 10.1016/0005-2736(81)90136-x

[72] Wood KC, Granger DN. Sickle cell disease: Role of reactive oxygen and nitrogen metabolites. Clinical and Experimental Pharmacology and Physiology. 2007;**34**:926-932. DOI: 10.1111/j.1440-1681.2007.04639.x

[73] Repka T, Hebbel RP. Hydroxyl radical formation by sickle erythrocyte membranes: Role of pathologic iron deposits and cytoplasmic reducing agents. Blood. 1991;**78**:2753-2758. DOI: 10.1182/blood.V78.10.2753.2753

[74] Walter PB, Fung EB, Killilea DW, Jiang Q, Hudes M, Madden J, et al. Oxidative stress and inflammation in iron-overloaded patients with beta-thalassaemia or sickle cell disease. British Journal of Haematology. 2006;**135**:254-263. DOI: 10.1111/j.1365-2141.2006.06277.x

[75] Calderon-Salinas JV, Munoz-Reyes EG, Guerrero-Romero JF, Rodriguez-Moran M, Bracho-Riquelme RL, Carrera-Gracia MA, et al. Eryptosis and oxidative damage in type 2 diabetic mellitus patients with chronic kidney disease. Molecular and Cellular Biochemistry. 2011;**357**:171-179. DOI: 10.1007/s11010-011-0887-1

[76] Maellaro E, Leoncini S, Moretti D, Bello BD, Tanganelli I, De Felice C, et al. Erythrocyte caspase-3 activation and oxidative imbalance in erythrocytes and in plasma of type 2 diabetic patients. Acta Diabetologica. 2011;**50**:489-495. DOI: 10.1007/s00592-011-0274-0

[77] Das BS, Nanda NK. Evidence for erythrocyte lipid peroxidation in acute falciparum malaria. Transactions of the Royal Society of Tropical Medicine and Hygiene. 1999;**93**:58-62. DOI: 10.1016/s0035-9203(99)90180-3

[78] Glenister FK, Coppel RL, Cowman AF, Mohandas N, Cooke BM. Contribution of parasite proteins to altered mechanical properties of malaria-infected red blood cells. Blood. 2002;**99**:1060-1063. DOI: 10.1182/blood.v99.3.1060

[79] Zoroglu SS, Armutcu F, Ozen S, Gurel A, Sivasli E, Yetkin O, et al. Increased oxidative stress and altered activities of erythrocyte free radical scavenging enzymes in autism. European Archives Psychiatry and Clinical Neuroscience. 2004;**254**:143-147. DOI: 10.1007/s00406-004-0456-7

[80] Yorbik O, Sayal A, Akay C, Akbiyik D, Sohmen T. Investigation of antioxidant enzymes in children with autistic disorder. Prostaglandins, Leukotrienes and Essential Fatty Acids. 2002;**67**:341-343. DOI: 10.1054/plef.2002.0439

[81] Altun H, Şahin N, Kurutaş EB, Karaaslan U, Sevgen FH, Fındıklı E. Assessment of malondialdehyde levels, superoxide dismutase, and catalase activity in children with autism spectrum disorders. Psychiatry and Clinical Psychopharmacology. 2018;**28**:408-415. DOI: 10.1080/24750573.2018.1470360

[82] Vos FE, Schollum JB, Coulter CV, Doyle TCA, Duffull SB, Walker RJ. Red blood cell survival in longterm dialysis patients. American Journal of Kidney

Diseases. 2011;**58**:591-598. DOI: 10.1053/j.ajkd.2011.03.031

[83] Nguyen-Khoa T, Massy ZA, de Bandt JP, Kebede M, Salama L, Lambrey G, et al. Oxidative stress and haemodialysis: Role of inflammation and duration of dialysis treatment. Nephrology Dialysis Transplantation. 2001;**16**:335-340

[84] Knap B, Prezelj M, Buturović-Ponikvar J, Ponikvar R, Bren AF. Antioxidant enzymes show adaptation to oxidative stress in athletes and increased stress in hemodialysis patients. Therapeutic Apheresis and Dialysis. 2009;**13**:300-305. DOI: 10.1111/j.1744-9987.2009.00728.x

[85] Shainkin-Kestenbaum R, Caruso C, Berlyne GM. Reduced superoxide dismutase activity in erythrocytes of dialysis patients: A possible factor in the etiology of uremic anemia. Nephron. 1990;**55**:251-253. DOI: 10.1159/000185970

[86] Gwozdzinski K, Pieniazek A, Gwozdzinski L. Reactive oxygen species and their involvement in red blood cell damage in chronic kidney disease. Oxidative Medicine and Cellular Longevity. 2021;**2021**:Article ID 6639199. DOI: 10.1155/2021/6639199

[87] Solas AB, Kutzsche S, Vinje M, Saugstad OD. Cerebral hypoxemia-ischemia and reoxygenation with 21% or 100% oxygen in new born piglets: Effects on extracellular levels of excitatory amino acids and microcirculation. Pediatric Critical Care Medicine. 2001;**2**:340-345. DOI: 10.1097/00130478-200110000-00011

[88] Dhinata KS. Common change of complete blood count parameters in COVID-19: A literature review. Journal of Medicine and Health. 2021;**3**:198-207. DOI: 10.28932/jmh.v3i2.3097

[89] Bartolini D, Stabile AM, Bastianelli S, Giustarini D, Pierucci S, Busti C, et al. SARS-CoV2 infection impairs the metabolism and redox function of cellular glutathione. Redox biology. 2021;**45**:102041. DOI: 10.1016/j.redox.2021.102041

[90] Kumar P, Osahon O, Vides DB, Hanania N, Minard CG, Sekhar RV. Severe glutathione deficiency, oxidative stress and oxidant damage in adults hospitalized with COVID-19: Implications for GlyNAC (glycine and N-acetylcysteine) supplementation. Antioxidants. 2021;**11**:50. DOI: 10.3390/antiox11010050

[91] Muhammad Y, Kani YA, Iliya S, Muhammad JB, Binji A, El-Fulaty Ahmad A, et al. Deficiency of antioxidants and increased oxidative stress in COVID-19 patients: A cross-sectional comparative study in Jigawa. Northwestern Nigeria. SAGE open medicine. 2021;**9**:2050312121991246. DOI: 10.1177/2050312121991246

[92] Mahdi A, Collado A, Tengbom J, Jiao T, Wodaje T, Johansson N, et al. Erythrocytes induce vascular dysfunction in COVID-19. Basic to Translational Science. 2022;**7**:193-204. DOI: 10.1016/j.jacbts.2021.12.003

[93] Thomas T, Stefanoni D, Dzieciatkowska M, Issaian A, Nemkov T, Hill RC, et al. Evidence of structural protein damage and membrane lipid remodeling in red blood cells from COVID-19 patients. Journal of Proteome Research. 2020;**19**:4455-4469. DOI: 10.1021/acs.jproteome.0c00606

[94] Young IS, Woodside JV. Antioxidants in health and disease. Journal of Clinical Pathology. 2001;**54**:176-186. DOI: 10.1136/jcp.54.3.176

[95] Masella R, Benedetto R, Vary R, Filesi C, Giovannini C. Novel

mechanisms of natural antioxidant compounds in biological systems: Involvement of glutathione and glutathione-related enzymes. The Journal of Nutritional Biochemistry. 2005;**16**:577-586. DOI: 10.1016/j.jnutbio.2005.05.013

[96] Nikolic-Kokic A, Blagojevic D, Spasic MB. Complexity of free radical metabolism in human erythrocytes. Journal of Medical Biochemistry. 2010;**29**:189-195. DOI: 10.2478/v10011-010-0018-7

[97] Chance B, Sies H, Boveris A. Hydroperoxide metabolism in mammalian organs. Physiological Reviews. 1979;**59**:527-605. DOI: 10.1152/physrev.1979.59.3.527

[98] Spasi MB. Antioxidative defence in mammals—A review. Jugoslovenska medicinska biohemija. 1993;**12**:1-9

[99] Kellogg EW, Fridovich I. Liposome oxidation and erythrocyte lysis by enzymically generated superoxide and hydrogen peroxide. The Journal of Biological Chemistry. 1977;**252**:6721-6728. DOI: doi.org/10.1016/S0021-9258(17)39909-X

[100] Gaetani GF, Galiano S, Canepa L, Ferraris AM, Kirkman HN. Catalase and glutathione peroxidase are equally active in detoxification of hydrogen peroxide in human erythrocytes. Blood. 1989;**73**:334-339

[101] Zińczuk J, Maciejczyk M, Zaręba K, Pryczynicz A, Dymicka-Piekarska V, Kamińska J, et al. Pro-oxidant enzymes, redox balance and oxidative damage to proteins, lipids and DNA in colorectal cancer tissue. Is oxidative stress dependent on tumour budding and inflammatory infiltration? Cancers. 2020;**12**:1636. DOI: 10.3390/cancers12061636

[102] Day BJ. Catalase and glutathione peroxidase mimics. Biochemical Pharmacology. 2009;**77**:285-296. DOI: 10.1016/j.bcp.2008.09.029

[103] Maciejczyk M, Zalewska A, Gryciuk M, Hodun K, Czuba M, Płoszczyca K, et al. Effect of normobaric hypoxia on alterations in redox homeostasis, nitrosative stress, inflammation, and lysosomal function following acute physical exercise. Oxidative Medicine and Cellular Longevity. 2022;**2022**:4048543. DOI: 10.1155/2022/4048543

[104] Fowkes SW. Antioxidant intervention in Down's syndrome. Smart Drug News. 1996;**4**:1-12

[105] Faut M, Paiz A, Martin S, de Viale LC, Mazzetti MB. Alterations of the redox state, pentose pathway and glutathione metabolism in an acute porphyria model. Their impact on heme pathway. Experimental Biology and Medicine. 2013;**238**:133-143. DOI: 10.1177/1535370212473702

[106] Pandolfi PP, Sonati F, Rivi R, Mason P, Grosveld F, Luzzatto L. Targeted disruption of the housekeeping gene encoding glucose 6-phosphate dehydrogenase (G6PD): G6PD is dispensable for pentose synthesis but essential for defense against oxidative stress. EMBO Journal. 1995;**14**:5209-5215. DOI: 10.1002/j.1460-2075.1995.tb00205.x

[107] Hattangadi SM, Lodish HF. Regulation of erythrocyte lifespan: Do reactive oxygen species set the clock? The Journal of Clinical Investigation. 2007;**117**:2075-2077. DOI: 10.1172/JCI32559

[108] Fujino T, Tada T, Beppu M, Kikugawa K. Purification and characterization of a serine protease in erythrocyte cytosol that is adherent to

oxidized membranes and preferentially degrades proteins modified by oxidation and glycation. The Journal of Biochemistry. 1998;**124**:1077-1085. DOI: 10.1093/oxfordjournals.jbchem.a022224

[109] Dubuisson M, Vander Stricht D, Clippe A, Etienne F, Nauser T, Kissner R, et al. Human peroxiredoxin 5 is a peroxynitrite reductase. FEBS Letters. 2004;**571**:161-165. DOI: 10.1016/j.febslet.2004.06.080

[110] Song Y, Tang L, Han J, Gao Y, Tang B, Shao M, et al. Uric acid provides protective role in red blood cells by antioxidant defense: A hypothetical analysis. Oxidative Medicine and Cellular Longevity. 2019;**2019**: Article ID: 3435174. DOI: 10.1155/2019/3435174

[111] Rizvi SI, Jha R, Maurya PK. Erythrocyte plasma membrane redox system in human aging. Rejuvenation Research. 2006;**9**:470-474. DOI: 10.1089/rej.2006.9.470

[112] Nishida Y. Inhibition of lipid peroxidation by methylated analogues of uric acid. Journal of Pharmacy and Pharmacology. 1991;**43**:885-887. DOI: 10.1111/j.2042-7158.1991.tb03204.x

[113] Revin VV, Gromova NV, Revina ES, Samonova AY, Tychkov AY, Bochkareva SS, et al. The influence of oxidative stress and natural antioxidants on morphometric parameters of red blood cells, the hemoglobin oxygen binding capacity, and the activity of antioxidant enzymes. BioMed Research International. 2019;**2019**. Article ID 2109269. DOI: 10.1155/2019/2109269

[114] Pandey KB, Rizvi SI. Biomarkers of oxidative stress in red blood cells. Biomedical Papers. 2011;**155**:131-136. DOI: 10.5507/bp.2011.027

[115] Luqman S, Rizvi SI. Protection of lipid peroxidation and carbonyl formation in proteins by capsaicin in human erythrocytes subjected to oxidative stress. Phytotherapy Research. 2006;**20**:303-306. DOI: 10.1002/ptr.1861

[116] Grinberg L, Newmark H, Kitrossky N, Rahamim E, Chevion M, Rachmilewitz E. Protective effects of tea polyphenols against oxidative damage to red blood cells. Biochemical Pharmacology. 1997;**54**:973-978. DOI: 10.1016/s0006-2952(97)00155-x

[117] Cesquini M, Torsoni M, Stoppa G, Ogo S. T-BOOH-induced oxidative damage in sickle red blood cells and the role of flavonoids. Biomedicine & Pharmacotherapy. 2003;**57**:124-129. DOI: 10.1016/s0753-3322(03)00018-0

[118] Tedesco I, Russo M, Russo P, Iacomino G, Russo GL, Carraturo A, et al. Antioxidant effect of red wine polyphenols on red blood cells. The Journal of Nutritional Biochemistry. 2000;**11**:114-119. DOI: 10.1016/S0955-2863(99)00080-7

[119] Blasa M, Candiracci M, Accorsi A, Piacentini M, Piatti E. Honey flavonoids as protection agents against oxidative damage to human red blood cells. Food Chemistry. 2007;**104**:1635-1640. DOI: 10.1016/j.foodchem.2007.03.014

[120] Sangai NP, Patel CN, Pandya HA. Ameliorative effects of quercetin against bisphenol A-caused oxidative stress in human erythrocytes: An in vitro and in silico study. Toxicology Research. 2018;**7**:1091-1099. DOI: 10.1039/c8tx00105g

[121] Morabito R, Remigante A, Marino A. Melatonin protects band 3 protein in human erythrocytes against H2O2-induced oxidative stress.

Molecules. 2019;**24**:2741. DOI: 10.3390/molecules24152741

[122] Veiko AG, Sekowski S, Lapshina EA, Wilczewska AZ, Markiewicz KH, Zamaraeva M, et al. Flavonoids modulate liposomal membrane structure, regulate mitochondrial membrane permeability and prevent erythrocyte oxidative damage. Biochimica et Biophysica Acta (BBA)—Biomembranes. 2020;**1862**:183442. DOI: 10.1016/j.bbamem.2020.183442

[123] Anjum R, Maheshwari N, Mahmood R. 3,4-Dihydroxybenzaldehyde mitigates fluoride-induced cytotoxicity and oxidative damage in human RBC. Journal of Trace Elements in Medicine and Biology. 2022;**69**:126888. DOI: 10.1016/j.jtemb.2021.126888

[124] Singh N, Rajini PS. Antioxidant-mediated protective effect of potato peel extract in erythrocytes against oxidative damage. Chemico-biological Interactions. 2008;**173**:97-104. DOI: 10.1016/j.cbi.2008.03.008

[125] Lins PG, Pugine SM, Scatolini AM, de Melo MP. In vitro antioxidant activity of olive leaf extract (*Olea europaea* L.) and its protective effect on oxidative damage in human erythrocytes. Heliyon. 2018;**4**(9):e00805. DOI: 10.1016/j.heliyon.2018.e00805

[126] Ghosh AK, Mitra EL, Dutta MO, Mukherjee DE, Basu AN, Firdaus SB, et al. Protective effect of aqueous bark extract of Terminalia arjuna on Cu^{2+} – ascorbate induced oxidative stress in vitro: Involvement of antioxidant mechanism(s). Asian Journal of Pharmaceutical and Clinical Research. 2013;**6**:196-200

[127] Magalhães AS, Silva BM, Pereira JA, Andrade PB, Valentão P, Carvalho M. Protective effect of quince (Cydonia oblonga Miller) fruit against oxidative hemolysis of human erythrocytes. Food and Chemical Toxicology. 2009;**47**:1372-1377. DOI: 10.1016/j.fct.2009.03.017

[128] Suwalsky M, Orellana P, Avello M, Villena F. Protective effect of Ugni molinae Turcz against oxidative damage of human erythrocytes. Food and Chemical Toxicology. 2007;**45**:130-135. DOI: 10.1016/j.fct.2006.08.010

[129] Hammami N, Athmouni K, Lahmar I, Ben Abdallah F, Belghith K. Antioxidant potential of Salicornia arabica lipid extract and their protective effect against cadmium induced oxidative stress in erythrocytes isolated from rats. Journal of Food Measurement and Characterization. 2019;**13**:2705-2712. DOI: 10.1007/s11694-019-00191-8

[130] Castro RI, Forero-Doria O, Soto-Cerda L, Peña-Neira A, Guzmán L. Protective effect of pitao (Pitavia punctata (R. & P.) Molina) polyphenols against the red blood cells lipoperoxidation and the in vitro LDL oxidation. Evidence-Based Complementary and Alternative Medicine. 2018;**2018**:1049234. DOI: 10.1155/2018/1049234

[131] Jawaid SA, Jain S, Bhatnagar M, Purkayastha S, Ghosal S, Avasthi AS. Free radical scavenging and antioxidant impact of Indian medicinal plant extracts on H2O2 mediated oxidative stress on human erythrocytes. American Journal of Phytomedicine and Clinical Therapeutics. 2014;**2**:1052-1069

Chapter 2

Erythrocytes as Messengers for Information and Energy Exchange between Cells

Erland Johansson and Anders B. Falk

Abstract

Evolution has created a hierarchy of systems for information and energy using different cells according to messages generated from DNA, RNA, and other sources. Erythrocytes are formed in high speed at about 2×10^6/s to balance dying or not working erythrocytes to maintain optimal energy and information transfer. Important information is handled by nucleotides and distribution of metal ions and phosphates when starting synthesis process. Handling of these processes needs kinases known to be magnesium-dependent. Oxygen delivered by erythrocytes is used by other cells to synthesize ATP and to increase reaction capacity. Complex signals to bone marrow balance erythroblasts before developing into reticulocytes and erythrocytes. We discuss some aspects of erythrocyte communication with other cells of the body with special focus on magnesium and selenium in this process.

Keywords: erythrocyte, magnesium, selenium, reactive oxygen species, glutathione, cholesterol, microRNA, kinase, hierarchy

1. Introduction

Erythrocytes provide oxygen and other necessary compounds to cells of the vascular compartment and other cells of the body. Erythrocytes also provide protection by collaborating in many ways with immune cells and humoral immunity [1, 2]. Erythrocytes communicate with other parts of the vascular compartment and to some extent also other parts of the body (**Figure 1**). Erythrocytes communicate as a transporter or by using signaling components perceived by receptors in other cells. We give some examples of how erythrocytes use magnesium and selenium in a hierarchy for optimal results as part of this communication. Magnesium occurs together with ATP, the energy currency of erythrocytes, and other cells. This makes magnesium necessary for many enzymes like the kinases of glycolysis. Further examples will be taken from lipids, glutathione, sphingosine-1-phosphate, purinergic signaling, RNA, the malaria parasite Plasmodium, and erythrocyte microvesicles.

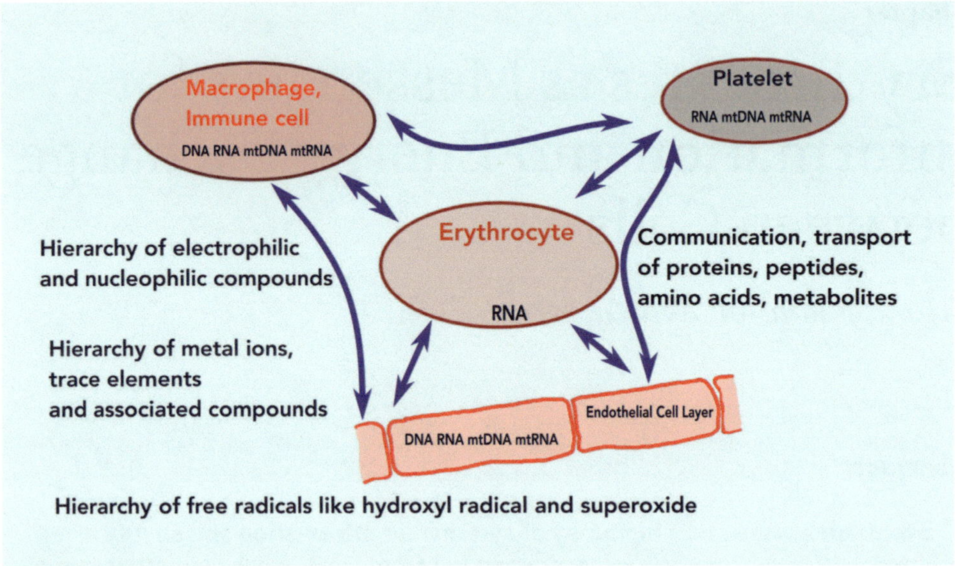

Figure 1.
Communication between erythrocytes and other cells takes place through RNA, proteins, metal ions, and other compounds that each form hierarchies. The first hierarchy involves free radical-induced production of compounds where hydroxyl radical and solvated electron reactions are involved. The second hierarchy involves electrophilic and nucleophilic compounds forming products adequate to the cells. The third hierarchy involves metal ions and ligands dependent on the previous two hierarchies but adapted to cell demand. In erythroid precursors and in other cells, excluding the erythrocyte, compounds and metal ions may reach DNA and RNA not controlled by evolutionary developed genome adapted to cell demand. Small changes of DNA and RNA, including epigenetic changes, may take place. These may become established or restored, but large damages will present symptoms and are difficult to restore.

2. Energy needs of the erythrocyte

Energy needs of the erythrocyte are fulfilled by ATP generated from glycolysis, which comprises several magnesium-dependent enzymes [3]. Magnesium is also necessary as enzyme cofactor or as part of enzyme substrate or product for more than 600 enzymes in the body, although not all these occur in the erythrocyte.

Among the enzymes of glycolysis, or the Embden-Meyerhof pathway, hexokinase, phosphofructokinase, phosphoglycerate kinase, and pyruvate kinase need magnesium as part of the ADP/ATP substrate or product. Aldolase and enolase need magnesium as a fundamental enzyme cofactor for stability and activity. The remaining glycolysis enzymes, glucose-phosphate isomerase, triosephosphate isomerase, glyceraldehyde 3-phosphate dehydrogenase, monophosphoglycerate kinase, and lactate dehydrogenase are not magnesium-dependent. Hexokinase, the first enzyme of glycolysis, increases reaction rate by at least 10 orders of magnitude compared to the uncatalyzed reaction, as an example of the very high catalytic demands on glycolytic enzymes [4]. Along with glycolysis, there is also the Rapoport-Luebering shunt producing 2,3-diphosphoglycerate, which is necessary for hemoglobin regulation. Magnesium level within erythrocytes is around 0.3 mM, which is three times higher than in plasma. The necessary levels of magnesium for glycolytic enzyme function are about an order of magnitude less than intra-erythrocyte magnesium concentration, suggesting that magnesium, while necessary for glycolysis, would not have a direct regulatory function in glycolysis [5]. Extracellular glucose can induce magnesium efflux in erythrocytes [6]. Erythrocytes of type II diabetes patients also have lower magnesium levels as compared to healthy individuals. Glycolysis is sometimes described as two competing mechanisms, the pentose-phosphate pathway and the Embden-Meyerhof pathway. The pentose-phosphate pathway, also called the hexose monophosphate shunt, is used by erythrocytes to generate reducing power in the form of NADPH and is not known to require magnesium. Glycolysis is regulated by positive or negative feed-back or feed-forward loops [7] and by the availability of oxygen. The pentose-phosphate shunt is favored when oxygen is abundant, whereas the Embden-Meyerhof pathway is favored in oxygen-limiting conditions. Erythrocyte glycolysis is also regulated by sphingosine-1-phosphate (S1P), at least in high altitudes [8], and by a circadian rhythm coupled to redox regulations [9].

Erythrocytes also contain insulin receptors at a copy number of about 1000–2000 per erythrocyte, depending on age and health status [10]. Erythrocyte insulin receptors respond to insulin and may regulate glycolysis, probably through phosphorylation of phosphofructokinase and intracellular redistribution of the enzyme [11]. This could be through a magnesium-dependent process, since insulin induces magnesium efflux from erythrocytes [12]. Signaling from the erythrocyte insulin receptor seems to be through the phospho-inositide pathway, since magnesium efflux was inhibited by wortmannin, a known inhibitor of phosphinositide-3-kinase [12]. Insulin receptor signaling also involves magnesium-dependent autophosphorylation of the tyrosine kinase part of the receptor. Erythrocytes are not dependent on insulin for glucose uptake, since they import glucose through the non-insulin-dependent glucose transporter GLUT-1. Insulin in synergism with the insulin C-peptide has been shown to inhibit the release of ATP from erythrocytes that occur as a result of low oxygen levels [13, 14]. This effect could be reversed by a phosphodiesterase 3 inhibitor [15]. Insulin and insulin receptors are also known to regulate the potassium balance of cells. This regulation is thought to mainly be at the expression of the genes encoding the voltage-gated potassium channels [16]. If so, then insulin and the insulin receptor are mainly at work in erythroid precursors for regulating potassium channels, rather than in the mature erythrocyte. High glucose concentration in plasma, also called hyperglycemia, as can be observed in diabetes, can have many effects on erythrocytes [17]. For instance, glucose can be metabolized in the erythrocyte by aldose reductase leading to sorbitol or fructose production through the polyol pathway, which can lead to complications like diabetic neuropathy [18]. Hyperglycemia also leads to glycation of hemoglobin. Glycated hemoglobin has an increased affinity for oxygen and could be expected to be less prone to release its oxygen in a normal way. However, the total oxygen delivery capacity of blood containing glycated hemoglobin is essentially unchanged [19]. Hyperglycemia can on the other hand cause cellular hypoxia by other mechanisms [20].

3. Glutathione and sphingosine-1-phosphate in the erythrocyte

Glutathione and sphingosine-1-phosphate are two compounds synthesized by and exported from erythrocytes. Glutathione is mainly a protector against oxidative damage, whereas sphingosine-1-phosphate is an immune cell communicator. Glutathione is synthesized by the enzymes glutamate-cysteine ligase and glutathione synthetase which are magnesium-dependent enzymes. Glutamate-cysteine ligase uses magnesium-ATP as substrate, and glutathione synthetase crystals show two magnesium ions in addition to ADP, glutathione, and sulfate [21]. Glutathione is the main reducing agent in erythrocytes and other cells. Supplementation with the glutathione prodrug N-acetylcysteine or glycine has been shown to improve several health aspects, including insulin resistance and cognition [22]. Erythrocytes and other tissues like liver and brain export glutathione to plasma [23]. Erythrocyte export of glutathione takes place by multi-drug resistance proteins [24]. Extracellular glutathione is of relevance in inflammation and disease [25–27] and can directly regulate immune components in plasma [28]. Intracellular glutathione concentration is usually three orders of magnitude higher than the extracellular glutathione concentration, and an energy-consuming active glutathione uptake mechanism seems not to be known [29]. Therefore, extracellular glutathione is not expected to be taken up by cells. Erythrocyte glutathione synthesis is dependent on the availability of the substrates glutamate and cysteine. Cysteine is imported as cystine through the glutamate/cystine antiporter, also known as SLC7A11 or system Xc-. Glutamate is acquired as glutamine through the ASCT2, also known as SLC1A5 transporter.

Sphingosine-1-phosphate (S1P) is a multifunctional molecule synthesized, stored, and exported to plasma by erythrocytes, platelets, and endothelial cells [30]. The erythrocyte is considered as the main contributor to plasma S1P. The two enzymes sphingosine-kinase-1 and sphingosine-kinase-2 are responsible for the synthesis of S1P. Targeted deletions of the genes coding for these enzymes suggest that erythrocytes and platelets are redundant for S1P synthesis under normal conditions, but necessary in systemic anaphylaxis [31]. Erythrocytes obtain sphingosine from plasma for S1P synthesis, although the precise import mechanism has not been elucidated. S1P is then synthesized by sphingosine-kinase-1, which is dependent on magnesium as shown by the presence of magnesium in the crystal structure of sphingosine-kinase-1 [32]. Evidence for intra-erythrocytic sphingosine synthesis has not been found [33]. The S1P-degrading enzymes sphingosine lyase and sphingosine phosphohydrolase are not present in erythrocytes, leading to some erythrocytic S1P storage capacity [34]. S1P is exported from erythrocytes by the major facilitator superfamily domain 2b (Mfsd2b), whereas export from endothelial cells is mediated through protein spinster homolog 2 (Spns2). S1P also needs apolipoprotein M in complex with high-density lipoprotein for effective export from erythrocytes [35].

S1P has several known effects in plasma. S1P forms a gradient where high levels are found peripherally and low levels are found within lymph nodes. Maturing lymphocytes expressing the S1P-receptor-1 proceed along this gradient, thereby leaving the lymph nodes. S1P1 receptor expression and further downstream signaling in endothelial cells are necessary for blood vessel integrity and vascular tone. Some of these effects are thought to be brought about by nitric oxide signaling [30]. Increased erythrocyte S1P levels have been found in COVID-19 patients [36]. S1P can reprogramme erythrocytes of chronic kidney disease patients to glucose metabolism through the Rapoport-Luebering shunt [37].

4. Erythrocytes, Alzheimer's disease, and other types of dementia

Several studies have indicated changes in erythrocytes of demented patients, including Alzheimer's disease. Binding of amyloid-beta to erythrocytes has been reported, and fibrils supposedly containing amyloid-beta have been visualized on erythrocytes [38]. A metabolomics study of Alzheimer's disease patients and controls identified 750 metabolites of which 7 increased and 24 decreased in erythrocytes of Alzheimer's disease patients [39]. The increased metabolites included argininate,

2-oxoarginine, and N-acetylarginine, all of which are known to form as a result of deficiency of the enzyme arginase. Erythrocyte arginase activity in Alzheimer's disease patients needs to be investigated to follow up on these results. All the 24 decreased metabolites were lipids, or related to lipid metabolism, whereof 10 were in the sphingolipid group like sphingomyelin. A metabolomic study of plasma in Alzheimer's disease similarly found differential presence of many lipids, but in this study sphingomyelin was higher in plasma of Alzheimer's disease patients [40]. Low blood hemoglobin levels and anemia have been associated, possibly in a causal relationship, with Alzheimer's disease and cognitive function [41].

Genome-wide association studies have found association between Alzheimer's disease and complement receptor 1 (CR1), clusterin, complement component 1 s (C1s), and in some ethnic groups also complement factor H [42–44]. CR1 is a transmembrane protein with a short cytoplasmic part, and many extracellular short consensus repeats (SCR) that bind complement component 3b (C3b) in complex with an antibody and its antigen, also referred to as an immune complex. The bound immune complex is then delivered to macrophages for internalization and degradation. CR1 has been localized to the membrane of erythrocytes where 80–90% of all CR1 in the body is estimated to be localized [45]. C3b binds to immune complexes with amyloid-beta, the main suspected protein in Alzheimer's disease. C3b-antibody-amyloid-beta complexes then bind to CR1 on erythrocytes that carry their load for delivery to the Kupffer macrophages in the liver. This process is sometimes referred to as immune adherence [46]. The anti-amyloid-beta antibody Aducanumab is a recent addition to Alzheimer's disease therapy. Aducanumab showed reduction of amyloid-beta in the brain, but no improvement of cognition or functional ability of patients [47]. Since antibodies have low ability to penetrate the blood-brain barrier, the effect of immunotherapy against Alzheimer's disease may involve amyloid-beta clearance from blood via erythrocytes and CR1 [48]. It should be noted that brain expression of CR1 is controversial, and therefore the erythrocyte-based explanation for CR1 involvement in Alzheimer's disease is preferred [45]. Complement has been shown to be active in normal brain function, for instance by performing synaptic pruning during development [49, 50]. The other complement components or complement factors that have been implicated in Alzheimer's disease may therefore be active in the brain. For instance, the C3b-binding complement factor H is present in both brain and plasma and has been shown to protect erythrocytes and other cells against complement-mediated damage [51]. As mentioned, complement factor H has shown association with Alzheimer's disease in some population studies [43]. A functional association between complement factor H and CR1 on erythrocytes in the pathology of Alzheimer's disease can therefore not be excluded. Inspection of CR1 alleles shows that CR1*2 is associated with increased risk for Alzheimer's disease. CR1*2 has an extra short consensus repeat in the extracellular domain of the protein. This length polymorphism is almost always associated with lower expression of CR1*2 on the erythrocyte surface [52]. The patients also show high levels of soluble CR1. The significance of soluble CR1 seems to be largely unknown. Clusterin, also known as apolipoprotein J, and ATP-binding cassette A7, also known as ABCA7, are two additional erythrocyte proteins [53, 54] that are known to be associated with Alzheimer's disease from genome-wide association studies.

Erythrocyte levels of the omega-3 fatty acids docosahexaenoic acid and eicosapentaenoic acid were lower in subjects with dementia in a longitudinal study population [55]. Possibly omega-3 fatty acids offer some protection against dementia as is further discussed in the section on cholesterol and lipids.

Erythrocytes have also been suggested to be a potential link between Alzheimer's disease and diabetes [56]. For instance, erythrocytes in Alzheimer's patients express more glucose transporter 1 (GLUT1) and insulin receptor than control subjects [57]. It is possible that these results are indications of a dysregulated metabolism that influences the availability of oxygen for the brain which might lead to cognitive impairment [58]. It should also be noted that Alzheimer's is more frequent in patients with type 2 diabetes.

5. Magnesium, cognition, and the erythrocyte

Body magnesium levels can be assessed through measurement of erythrocyte or plasma magnesium levels. Since only about 1% of total-body magnesium is found in blood, further tests may be necessary for more thorough evaluation of whole-body magnesium levels. A correlation between intra-erythrocyte magnesium levels and cognition was found in rats [59], and a similar association could be found in a human study involving patients with vascular cognitive impairment [60]. Erythrocytes show significantly diminished intracellular magnesium levels in patients with vascular cognitive impairment, although plasma magnesium levels are normal. This could be interpreted as a measure of whole-body magnesium levels, including the brain. Low magnesium levels in the central nervous system are known to be associated with complications like depression. An explanation for this may be the excitotoxicity caused by the *N*-methyl-D-aspartate (NMDA) receptor, a glutamate-regulated calcium channel present in the plasma membrane of nerve cells. The NMDA receptor requires magnesium as a "gate-keeper" to prevent opening of the channel. Small areas of blood-brain contact can occur in dementia as microbleeds or microhemorrhages, also known as blood extravasations [61]. Communication between brain and the immune complement and blood coagulation systems has been suggested to be part of the pathology in vascular dementia [62]. A meta-analysis of dementia trials showed that vascular dementia was associated with increased levels of fibrinogen, activated factor VII, factor VIII, von Willebrand factor, D-dimer, and homocysteine [63]. When displaying phosphatidylserine on the surface, erythrocytes contribute to coagulation by interacting with the gamma-carboxyglutamyl (Gla) domains on coagulation factors, initiating the formation of thrombin from prothrombin. Phosphatidylserine exposure is regulated by intracellular calcium level and by the enzymes flippase and scramblase. Flippase moves phospholipids from the exoplasmic to the cytoplasmic side of the plasma membrane. Flippase activity in human erythrocytes is performed by ATP11C, a P4-ATPase [64] that has been crystallized together with its interacting protein CDC50, which is also known as TMEM30A [65]. The structure is similar to other human or yeast P4-ATPase flippase structures [66, 67]. Reduction in ATP11C activity leads to increased phosphatidylserine exposure on erythrocytes [68] and in case of deficiency may lead to mild hemolytic anemia [69]. Flippase activity is dependent on ATP and magnesium, and some of the 3D structures also show magnesium ions at the phosphorylation site [66, 67]. Lower magnesium levels in erythrocytes can therefore lead to lower flippase activity and more exoplasmic exposure of phosphatidylserine.

Scramblase moves lipids between the two monolayers of a membrane in either direction, thus evening out differences in lipid composition between the two sides. Human scramblase comes in three different protein families [70]. The scramblase activity of erythrocytes is performed by transmembrane protein 16F (TMEM16F, also known as anoctamin6) [71, 72]. TMEM16F is a calcium-dependent homodimeric structure with 10 transmembrane alfa-helices and a large amino-terminal cytosolic domain in each subunit [73]. The hydrophilic head of the lipid substrate proceeds in a cavity on TMEM16F, similar to the swiping of a credit card [74]. Evidence for TMEM16F scramblase activity also comes from patients with Scott syndrome [75, 76], a condition involving low coagulation ability of platelets and erythrocytes [77]. Three calcium-binding sites can be seen in TMEM16F high-resolution structures, but so far, crystal structures of the TMEM family seem not to have included magnesium. Since TMEM16F is negatively regulated by magnesium [78], low intracellular magnesium levels, as was found in erythrocytes of vascular dementia patients, may lead to more TMEM16F scramblase activity, phosphatidylserine exposure, and blood coagulation. Magnesium binding to TMEM proteins could involve a magnesium-calcium competition similar to some other calcium-dependent proteins often involving the regulatory protein calmodulin [79]. A direct interaction between TMEM proteins and calmodulin has been suggested but is still controversial [80].

6. Iron and the erythrocyte

Iron is necessary for oxygen transport by hemoglobin in the erythrocyte. In addition, iron is stored in ferritin, which has been shown to occur in the erythrocyte and may be necessary for binding excess erythrocyte iron. Excess erythrocyte iron may be a consequence of hemoglobin oxidation and degradation and may be a normal part of erythrocyte aging. The erythrocyte also contains the iron export protein ferroportin, which may likewise protect the erythrocyte from toxic effects of excess iron. Ferroportin is regulated by the iron hormone hepcidin, which in turn is regulated by interleukin-6. Conditional deletion of ferroportin in mice leads to build-up of intracellular erythrocyte iron and may result in hemolysis [81]. The ferroportin Q248H mutation protects ferroportin from degradation caused by hepcidin and seems to have been selected in African populations possibly due to some protection against malaria conferred by the mutation. Ferroportin functions in erythrocytes and erythroid cells as an export gate for the regulation of total-body iron homeostasis [82, 83]. Iron is dysregulated in ferroptosis, a form of regulated cell death distinct from apoptosis [84]. Ferroptosis is characterized by iron overload, peroxidation of lipids and low activity of the selenium-containing enzyme glutathione peroxidase 4. Ferroptosis has been associated with several pathological conditions including neurological diseases such as Alzheimer's disease [85]. Transcriptomic analysis reveals changed expression levels of many ferroptosis-related genes in Alzheimer's disease patients [86], including the selenium-containing enzyme glutathione peroxidase 4. Neuron ferroportin needs the amyloid precursor protein for stability and localization. Degradation of amyloid precursor protein to amyloid-beta may lead to increased iron levels in neurons and pave the way for ferroptosis-induced neuron death [87]. Some amyloid-beta will reach the vascular compartment by way of the glymphatic system. Amyloid-beta can then bind to erythrocytes in the vascular compartment. Amyloid-beta has been shown to induce morphological changes in erythrocytes [88], affect signal transduction [89], and inhibit production and release of ATP from erythrocytes [90]. Amyloid-beta could potentially affect erythrocytes leading to dysregulation of total-body iron homeostasis, potentially worsening the prodromal phase of Alzheimer's disease, although this remains to be tested.

Hereditary hemochromatosis is another condition also characterized by iron overload in several tissues of the body. Particularly the liver, heart, pancreas, and skin are affected in hereditary hemochromatosis. Erythrocytes of patients with hereditary hemochromatosis may show morphological changes [91], and the patients can be affected by secondary diseases that affect erythrocytes, such as polycythemia vera or hemolytic anemia [92].

7. The erythrocyte and purinergic signaling

Cells of the vascular compartment communicate with each other by adenosine and its nucleotides ADP and ATP, a process called purinergic signaling. A set of 19 receptor subunits form receptors for adenosine and its nucleotides [93]. The ATP-binding P2X7 receptor is found on erythrocytes and many other cells in the vascular compartment. Binding of the ATP ligand to the P2X7 receptor of erythrocytes leads to phosphatidylserine exposure and clearance by macrophages [2]. ATP generated from erythrocytes can activate the P2X7 receptor on other cells like endothelial and myocardial cells. This can lead to inflammasome activation and may lead to pyroptosis of the cell through activation of gasdermin [94]. Structural studies of the P2X7 receptor show a "leaping dolphin"-like structure, where the head is formed by the extracellular domain and the tail by the transmembrane helices [95]. Unlike other P2X receptors, the P2X7 receptor shows a rather extended C-terminal domain described as the "cytoplasmic ballast" [96]. The cytoplasmic ballast contains one GDP-binding site, two zinc-binding sites and similarity to TNF receptor I and lipopolysaccharide binding domains [97]. When exposed to high ATP concentrations, the P2X7 receptor can form a macropore that allows passage of solutes up to 900 Da in size leading to apoptosis of the cell. The macropore size is sometimes incorrectly stated to be 900 kDa [94, 98, 99]. The macropore may be formed by the P2X7 receptor or together with some other membrane proteins like pannexin or connexin. Some evidence suggests that the P2X7 receptor is inhibited by magnesium [100]. The P2Y12 receptor is expressed on erythrocytes but is most known for its expression on platelets, where binding to ADP is part of the process of platelet activation, which is an important part of blood coagulation. Due to this, the P2Y12 receptor has been the focus of pharmaceutical development for instance leading to the antagonists clopidogrel, prasugrel, and ticagrelor. The P2Y13 receptor is activated by ADP on erythrocytes leading to diminishing of ATP export. Purinergic signaling in erythrocytes is part of the interaction between erythrocytes, platelets, and endothelial cells, both in normal physiological conditions and in pathological conditions such as diabetes [101].

Export of ATP from erythrocytes takes place through the pannexin transmembrane protein. ATP can then be metabolized in the extracellular space, for instance in plasma, to ADP and AMP by ectonucleoside triphosphate diphosphohydrolase, usually called CD39. AMP can then be further metabolized to adenosine by ecto-5′-nucleotidase, usually called CD73. CD39 and CD73 are transmembrane proteins expressed on the plasma membrane of many cell types including endothelial cells [102, 103]. Crystal structures of CD39 and CD73 show calcium and zinc ions [104, 105]. Interestingly, CD39 and CD73 collaborate with heme oxygenase-1 in heme catabolism [106]. Adenosine generated by CD73 increases heme oxygenase-1 in macrophages through stimulation of the adenosine A2A or A2B receptors. Heme and ATP can be generated from erythrocytes as a consequence of hemolysis. The intracellular pool of adenosine can be replenished by equilibrative nucleoside transporter 1 [101], which is a transporter localized in the plasma membrane of erythrocytes and other cells. AMP can then be regenerated in the erythrocyte from adenosine by adenosine kinase [107], an enzyme in the purine nucleotide salvage pathway [108].

Uric acid, the result of adenosine and guanosine catabolism, is found in plasma where it may have antioxidative properties [109]. Significant correlations between uric acid levels and several erythrocyte parameters have been found, such as mean corpuscular volume, mean corpuscular hemoglobin concentration, and erythrocyte distribution width [109, 110]. However, decreased plasma uric acid levels accomplished by the recombinant urate oxidase Pegloticase did not affect the oxidative status of plasma [111]. A possible interpretation of this result is that the antioxidative properties of erythrocytes could compensate for the loss of uric acid's antioxidative capacity.

8. The erythrocyte, cytokine, and immune cell relation

Erythrocytes have been found to be a reservoir of cytokines, a group of immune signaling proteins. Cytokines include chemokines, interferons, interleukins, and the hormone erythropoietin. Close to 50 different cytokines have been identified in or associated with erythrocytes [112]. Erythrocytes probably due to cytokine storage have a role in defense against pathogens, immune function, and homeostasis [2]. Erythrocytes also interact with the cellular part of the immune system (**Figure 1**). Macrophages are an important part of the cellular immune system that participate in both the birth and death of erythrocytes. Macrophages phagocytose senescent erythrocytes when they pass through liver or spleen. A phagocytosis signal is provided by phosphatidylserine exposure on erythrocytes, recognized by the TIM (T cell immunoglobulin and mucin domain containing) and CD300 receptors on macrophages. The CD47 membrane protein, which is present on erythrocytes, binds to the SIRPalfa protein on macrophages to downregulate phagocytosis [113]. Erythrocytes stored for long time periods can induce M2 macrophage polarization through the immunosuppressive interleukin-10 [114], thereby downregulating immunity. Macrophages are also necessary in erythropoiesis, the formation of new erythrocytes, where they interact with the erythroid progenitors in erythroblastic islands [115, 116]. Macrophages promote erythroblast proliferation and differentiation by secreting growth factors, providing nutrients like iron and finally phagocytosing nuclei of the nascent reticulocytes. Early-stage erythroblasts respond to growth factors like interleukin-3, stem cell factor, and erythropoietin. Several receptor-ligand pairs facilitate macrophage-erythroblast interactions. Vascular cell adhesion molecule1 of the macrophage interacts with integrin alfa4beta1 of the erythroblast and integrin alfaV of the macrophage interacts with intercellular adhesion molecule4 of the erythroblast [115]. Erythroblast macrophage protein of the macrophage interacts with erythroblast macrophage protein expressed on the surface of the erythroblast. CD163 and CD169 are expressed on macrophages and are known to be necessary for erythropoiesis, but the corresponding molecule on the erythroblast has not been identified. Finally, DNase2alfa of the macrophage is necessary for the phagocytosing of the nuclei of the nascent reticulocytes [116]. Selenium and selenoproteins are other factors necessary for erythropoiesis. Mutation of the selenocysteine-transfer-RNA, sometimes abbreviated Trsp, selenoprotein W, and glutathione peroxidase 4 genes in mice led to defective erythropoiesis [117–119]. Selenoprotein W has been suggested to be an adaptor protein to the E3 ubiquitin ligase TRIM21 [120]. Ubiquitin is a protein that delivers proteins for degradation to the proteasome, a protein complex that degrades mainly damaged proteins marked by ubiquitin. The 20S proteasome and ubiquitin have been detected in reticulocytes and mature erythrocytes. Erythrocytes from patients with Alzheimer's or Parkinson's disease show decreased 20S proteasome activity [121, 122].

Erythrocytes also interact with other cells of the immune system. Erythrocytes have been shown to inhibit T cell activation or activation-induced apoptosis presumably via reactive oxygen species-dependent pathways [123, 124]. Erythrocytes treated with a cancer cell line stimulated T cells to more proliferation and other cytokine secretion profile than if treated with control erythrocytes [125]. The C-C chemokine RANTES (regulated on activation, normal, T cell expressed, and secreted) guides transendothelial migration of eosinophils, an immune cell that is responsible for interleukin-5 production and largely involved in asthma and allergy. Erythrocytes regulate this process by scavenging RANTES in the vascular compartment [126].

9. Erythrocytes and function of the vascular compartment

The erythrocyte is important for several aspects of the vascular compartment, in particular vascular tone and vascular integrity. One aspect is the signaling based on nitric oxide, a gaseous molecule produced in the vascular compartment by erythrocytes, platelets, and endothelial cells by endothelial nitric oxide synthase (eNOS). Erythrocytes both produce and release nitric oxide [127] but can also scavenge nitric oxide by hemoglobin [128]. The synthesis of nitric oxide proceeds from arginine and oxygen-generating nitric oxide and citrulline. Regulation of nitric oxide synthesis is performed by arginase I by degrading the substrate arginine to ornithine and urea. Nitric oxide synthesis can also take place by deoxy-hemoglobin acting as nitrite reductase in hypoxia.

Nitric oxide is a vasodilator based on its effects on the vascular smooth muscle cells that surround the vascular compartment and contribute to vascular tone. The vascular smooth muscle cells contain the nitric oxide receptor soluble guanylate cyclase that produces cyclic guanosine monophosphate (cGMP). Protein kinase G is a further downstream signaling component, but the steps leading to vasodilation seem to be unknown in exact detail [129], although the last step probably involves changes in myosin phosphorylation, performed by Rho-associated kinase, zipper interacting kinase [130], myosin light chain kinase, or myosin phosphatase [131]. Some of these enzymes are also regulated by calmodulin and may therefore also be indirectly regulated by magnesium [79]. Soluble guanylate cyclase and protein kinase G are also present in erythrocytes and platelets, although the precise function of this pathway in erythrocytes is unknown [132]. Nitric oxide has an inhibitory effect on platelet activation, possibly through phosphorylation of the thromboxane receptor [133]. The effects of nitric oxide also include S-nitrosation of proteins and regulation of cGMP-gated ion channels by cGMP [128]. For instance, nitric oxide protects against myocardial infarction through S-nitrosation of the mitochondrial permeability transition pore regulator cyclophilin D [134]. Nitrosation of cysteine-93 of the beta-chain of hemoglobin has been suggested as a transport mechanism for nitric oxide in the erythrocyte. This hypothesis is not currently favored as a result of mutation studies [127]. A magnesium-dependent nitrosation of glutathione, yielding S-nitrosoglutathione, has been reported and could potentially occur in erythrocytes [135]. The physiological significance of S-nitrososglutathione is however unclear.

The relative contribution of different cell types to nitric oxide production in the vascular compartment is not yet fully elucidated. Erythrocytes both produce and scavenge nitric oxide, which complicates the interpretation their contribution. Several evidences from pathological conditions point to the importance of erythrocytes for generation of nitric oxide bioactivity. Increased expression of erythrocyte arginase in diabetes leads to less nitric oxide production from erythrocytes. ENOS can be monomerized in erythrocytes of type 2 diabetes patients, usually referred to as uncoupling. The monomers then produce superoxide, which is also produced by nicontinamide adenine dinucleotide oxidase (NOX), both in erythrocytes and in endothelial cells. This results in dysfunction of endothelial cells [136], a common phenomenon in type 2 diabetes. As a further example, studies of anemic patients and a mouse model of anemia show that anemia is associated with erythrocyte dysfunction and reduced nitric oxide bioactivity. ENOS activity is then increased in the vascular wall and heart as compensatory mechanisms. If anemia is combined with endothelial dysfunction, the compensatory nitric oxide bioactivity may not be sufficient and could lead to adverse outcomes in myocardial infarction [137, 138].

10. Erythrocytes, cholesterol, and other lipids

The erythrocyte membrane contains lipids like phosphatidylserine, phosphatidylethanolamine, phosphatidylcholine, sphingomyelin, and membrane proteins. The membrane also contains cholesterol, which is important for membrane stability and deformability of erythrocytes [139]. The parasite Plasmodium does not synthesize cholesterol, and vesicles containing cholesterol have been observed to transport from erythrocytes to the Plasmodium parasite [140]. Cholesterol can be chelated by beta-methyl-cyclodextrin, which has been used against Plasmodium infections [141]. Many studies point to an association between lipid constituents of erythrocytes and divers diseases [142]. Many studies have found that the omega-3, also called n-3 fatty acids, are promoting health in various ways, whereas omega-6 (n-6) may be less health promoting and the saturated fatty acids are further less health promoting. Trans-fatty acids are usually not considered health promoting, although this is somewhat controversial considering their occurrence in dairy products. Similar results are found across many studies, irrespective of whether plasma or erythrocyte levels are measured. One reason for less clear health promotion of omega-6 fatty acids is that the omega-6 arachidonic acid is a precursor for pro-inflammatory prostaglandins, thromboxanes, and leukotrienes [143], collectively called eicosanoids. Omega-3 fatty acids on the other hand have been reported to reduce signaling through the pro-inflammatory toll-like receptor 4 (TLR4) [144]. Dietary levels of omega-6 fatty acids are usually well in excess of those of omega-3, which may lead to undesirable enzymatic competition for the fatty acid substrate. Erythrocyte membrane constituents are of interest as diagnostics, since erythrocytes are more long-term indicators than plasma levels, which are influenced by the most recent meal. Erythrocyte levels of omega-3 fatty acids were associated with less cardiovascular events in the Framingham heart study [145]. Erythrocyte omega-6 fatty acid levels showed no association with cardiovascular events in the same study [146]. Erythrocyte omega-3 polyunsaturated fatty acids were negatively correlated with cancer [147]. Erythrocyte omega-3 fatty acids were associated with less risk of islet autoimmunity in children with diabetes [148]. A study of erythrocytes in preclinical Alzheimer's disease found increased levels of the omega-6 fatty acid arachidonic acid and decreased levels of the omega-3 fatty acid docosapentaenoic acid in participants with high neocortical beta-amyloid load prior to cognitive impairment [149]. Decreased omega-3 fatty acid levels in erythrocyte membranes of schizophrenia patients has given rise to a modified version of the dopamine hypothesis of schizophrenia [150]. A study of erythrocyte fatty acids in schizophrenia patients found that omega-3 polyunsaturated fatty acids were significantly lower and the omega-6:omega-3 fatty acid ratio was significantly higher in the group with dominantly negative symptoms as compared to the group with dominantly positive symptoms [151]. A significant deficit of erythrocyte omega-3 docosahexaenoic acid has been noted in bipolar disorder type I [152], whereas erythrocyte fatty acid profiles had no predictive value in autism spectrum disorder [153].

Trans-fatty acids in erythrocytes were found to be associated with increased risk of coronary heart disease in prospective [154] or cross-sectional studies [155]. However, trans-fatty acids in erythrocytes appear to be decreasing in Europe as seen in samples from 2008 to 2015 [156].

11. Erythrocytes and ribonucleic acid (RNA)

Although ribonucleic acid (RNA) has been reported in erythrocytes since long (**Figure 1**), the purity of blood cell preparations was often questioned [157]. The use of single-cell transcriptomics has provided further confirmation of the presence of RNA in erythrocytes [158]. Around 8000 messenger-RNA transcripts have now been found in erythrocytes. Analysis of messenger-RNA has shown that erythrocytes can be divided into at least seven different categories [158]. These seem to represent different stages of development from reticulocytes to cells in transition to mature and finally senescing erythrocytes. The most abundant messenger-RNA transcripts are related to erythropoiesis and may be seen as a residue from that process. Alternatively, messenger-RNA may serve as template for protein synthesis, which would require the presence of ribosomes and other parts of translation, which may possibly exist at a low level in erythrocytes [159]. Hundreds of different microRNAs (miRNAs) have also been confirmed in erythrocytes [159]. MiRNAs are usually about 22 nucleotides in length and function to reduce, or silence, the expression of genes. Important parts of this mechanism are the RNAse III enzymes Drosha and Dicer, the endonuclease Argonaute, and the RNA-binding protein DGCR8. Drosha, Dicer, and Argonaute are all magnesium-dependent enzymes [160–162]. MiRNA genes are transcribed by RNA polymerase II to a pri-miRNA. Drosha and DGCR8 then process the pri-miRNA to a pre-miRNA which is exported from the nucleus through exportin-5. Once in the cytoplasm, the miRNA is further processed by Dicer and bound to Argonaute2 forming the RNA-induced silencing (RISC) complex. Some of the miRNAs of the erythrocyte like miR-451, miR-144, and miR-486 are involved in erythropoiesis [163]. MiR-451 is also associated with malaria, and miR-144 is correlated with hypoxia at high altitudes [159]. An attractive hypothesis is that erythrocytic RNA enclosed in microvesicles is used as a means of communication between erythrocytes and other cells.

Microvesicles are extracellular vesicles 0.1–1 micrometer in size, formed by budding from the plasma membrane. Microvesicles are distinguished from exosomes, which are smaller and released as preformed vesicles, 10–100 nanometer in size. Microvesicles and exosomes may contain proteins, lipids, and RNA. Microvesicles are particularly known as carriers of miRNA. Erythrocytes are known to form microvesicles under blood storage, but microvesicles are also formed as a normal physiological process [164]. As mentioned, microvesicles may fulfill a function as information carriers between different cells in the body. Microvesicles also function as an efficient protection against proteases or RNases that would otherwise degrade the content [165]. A connection between erythrocyte microvesicle formation and diseases or pathological conditions have been proposed in several cases. However, a more comprehensive analysis of miRNA content of microvesicles derived from healthy erythrocytes compared to those from patients with disease conditions seems to be needed.

Transport of miRNA or messenger-RNA from erythrocytes to other cells may be a way of regulating gene expression in target cells [159]. The known functions of these miRNAs in other cells of the body could accordingly be relevant also in the context of erythrocytes. MiRNAs are generally edited by the gene silencing mechanism in cells, giving rise to several variants of each miRNA family, often indicated in the nomenclature by extra suffixes, like the expressed strand being indicated by a 3-p or 5-p suffix. Some abundant and notable erythrocyte miRNAs are miR451, miR144, miR16, and let-7. The miR451 family is abundant in erythrocytes and functions in erythropoiesis by downregulating the Ywhaz gene, whose protein product 14–3-3-zeta keeps the transcription factor FoxO3 in the cytosol [166]. FoxO3 positively regulates antioxidant enzymes like catalase and glutathione peroxidase. CRISPR-Cas9-mediated mutagenesis of miR451 confirmed that miR451 is necessary for erythroid differentiation and expression of transferrin receptor 1, also known as CD71 [163]. The miR144 family is expressed together with miR451 and was similarly found to be necessary for erythroid differentiation and CD71 expression [163]. Plasmodium-infected erythrocytes produce microvesicles containing miR451, miR16, and let-7, among others [167–169].

12. Erythrocyte and other organisms, particularly plasmodium

Erythrocytes interact with several organisms like bacteria and Plasmodium. Erythrocytes also interact with viruses, although mainly through adherence of viruses to erythrocyte membrane proteins or glycocalyx. Several erythrocyte membrane proteins are known as receptors for Plasmodium, somewhat dependent on the Plasmodium species. An interesting aspect is that Plasmodium seems to be dependent on iron acquired directly from hemoglobin of the erythrocyte. For instance, Plasmodium cannot obtain iron from heme. The erythrocyte is made dependent on the Plasmodium parasite by the glutathione synthesis of the parasite, whereas the erythrocytes own glutathione synthesis seems to be largely turned off [170]. Glutathione export from the Plasmodium-infected erythrocyte proceeds mainly in the oxidized (GSSG) form probably reflecting conditions of oxidative stress under infection. Erythrocytes infected with Plasmodium produce microvesicles that contain miR451, let7, and Argonaute2 protein together forming a functional RISC complex [167]. Plasmodium does not contain the genes necessary for miRNA production, but human miRNA and Argonaute protein have been detected in Plasmodium, presumably transferred from the erythrocyte by microvesicles. Infected erythrocytes also produce microvesicles that can be transferred to endothelial cells, astrocytes, and microglia [171]. Plasmodium causes profound changes in the erythrocyte plasma membrane, which can become almost devoid of cholesterol and display phosphatidylserine [172]. Absence of cholesterol makes the plasma membrane susceptible to pore formation by granulysin, a pore-forming peptide produced by cytotoxic cells like gamma-delta T cells. Exoplasmic phosphatidylserine may make the cells less susceptible to perforin, another pore-forming peptide. Infected erythrocytes display PfEMP (Plasmodium falciparum erythrocyte membrane protein) which binds to receptors on endothelial cells and thereby prevents circulation of infected erythrocytes. Although they may display phosphatidylserine, infected erythrocytes are consequently not easily cleared from circulation by spleen macrophages. PfEMP can be recognized by gamma-delta T cells of the host, but the parasite can change the expressed PfEMP, since it contains about 60 different PfEMPs. Although malaria is usually not fatal, cerebral malaria caused by Plasmodium falciparum can be fatal particularly in children. Cerebral malaria involves a partial disruption of the blood-brain barrier [173]. This probably comes about through phagocytosis of infected erythrocytes by endothelial cells in brain capillaries, followed by presentation of Plasmodium antigens on the endothelial cell and attack from alfa-beta Tcells [173].

13. Medical and societal impact of erythrocyte science

Treatments and diagnosis based on erythrocytes have been much sought for, and many have been developed. For instance, several modulators of S1P-receptors have been developed, the first and most well-known being fingolimod. These are now being used as therapy for multiple sclerosis. Fingolimod is first phosphorylated to fingolimod-phosphate by sphingosine-kinase-2, which resides in the nucleus of nucleated cells. The S1P-receptor-1 is then internalized, thus preventing lymphocyte egress along the S1P gradient. The result is reduced damage from the immune system on the central nervous system or other organs. It is likely that new treatments or modulators based on S1P will be developed.

Erythrocyte microvesicles present diagnostic and possibly also therapeutic opportunities. Erythrocyte microvesicles may be used as carriers in pharmaceutical applications, or as alternatives to whole cells in cell therapy applications. Microvesicles are less complex to handle in biological product formulation and production and confer less safety issues than cells [174]. Erythrocyte microvesicles could be used to deliver the miRNAs they harbor naturally or engineered for more specific delivery. Potential challenges include multiple effects of the same miRNA and the existence of many different miRNAs, with different and opposing effects, in the same microvesicle. It is also important to consider the conditions that favor microvesicle formation. Lysophosphatidic acid (LPA) has been shown to induce phosphatidylserine exposure and microvesicle formation [175].

Malaria is the disease caused by the erythrocyte parasite Plasmodium. Malaria is still of considerable importance as a devastating disease in many parts of the world. It is to be expected that new knowledge on erythrocytes and the interaction with Plasmodium could lead to new treatments against malaria. For instance, it may be possible to interfere with the attachment of infected erythrocytes to the endothelium [176]. That strategy may circumvent undesirable side effects of many other approved antimalarial drugs.

Population levels of selenium can be increased by fortification of foodstuff [177] or by fertilization of crops. Foliar selenium fertilization has been shown to increase selenium levels in serum and erythrocyte glutathione peroxidase activity in subjects consuming the crop [178]. That selenium in foodstuff finds it way to the erythrocyte was elegantly shown in the case of selenoneine from the beluga whale that was found in erythrocytes of inuits [179]. Magnesium can be easily replenished by mineral supplements, or by a diet rich in vegetables and some other known sources of magnesium. More reliable diagnostics for assessment of whole-body magnesium status may be desirable to detect subclinical cases of magnesium deficiency. Several of the diseases or nutritional imbalances mentioned are of substantial importance, especially considering growing human populations, faster human communications, and climate change. All this may contribute to a changed exposure to pathogens like bacteria, virus, and Plasmodium. Erythrocytes as biomarkers of changed element profiles may provide one way to observe early deviations and provide customized diet suggestions and earlier treatment of diseases.

Author details

Erland Johansson* and Anders B. Falk
EJSelenkonsult AB, Uppsala, Sweden

*Address all correspondence to: 101jejohansson@gmail.com

References

[1] Anderson HL, Brodsky IE, Mangalmurti NS. The evolving erythrocyte: Red blood cells as modulators of innate immunity. Journal of Immunology. 2018;**201**(5):1343-1351. DOI: 10.4049/jimmunol.1800565

[2] Johansson E. In: Fatima Shad K, editor. B. Falk a. Erythrocytes as Biomarkers of Virus and Bacteria in View of Metal Ion Homeostasis, Erythrocyte—A Peripheral Biomarker For Infection and Inflammation. Rejika: In, IntechOpen; 2021. DOI: 10.5772/intechopen.97850

[3] van Wijk R, van Solinge WW. The energy-less red blood cell is lost: Erythrocyte enzyme abnormalities of glycolysis. Blood. 2005;**106**(13):4034-4042. DOI: 10.1182/blood-2005-04-1622

[4] Koshland DE. Molecular geometry in enzyme action. Journal of Cellular and Comparative Physiology. 1956;**47**(S1): 217-234. DOI: 10.1002/jcp.1030470415

[5] Laughlin MR, Thompson D. The regulatory role for magnesium in glycolytic flux of the human erythrocyte. Journal of Biological Chemistry. 1996;**271**(46):28977-28983. DOI: 10.1074/jbc.271.46.28977

[6] Ferreira A, Rivera A, Wohlgemuth JG, Dlott JS, Snyder LM, Alper SL, et al. Dysregulated erythroid Mg^{2+} efflux in type 2 diabetes. Frontiers in Cell and Development Biology. 2022;**10**:861644. DOI: 10.3389/fcell.2022.861644

[7] Locasale JW. New concepts in feedback regulation of glucose metabolism. Current Opinion in Systems Biology. 2018;**8**:32-38. DOI: 10.1016/j.coisb.2017.11.005

[8] Sun K, Zhang Y, D'Alessandro A, Nemkov T, Song A, Wu H, et al. Sphingosine-1-phosphate promotes erythrocyte glycolysis and oxygen release for adaptation to high-altitude hypoxia. Nature Communications. 2016;**7**(1):12086. DOI: 10.1038/ncomms12086

[9] Ch R, Rey G, Ray S, Jha PK, Driscoll PC, Dos Santos MS, et al. Rhythmic glucose metabolism regulates the redox circadian clockwork in human red blood cells. Nature Communications. 2021;**12**(1):377. DOI: 10.1038/s41467-020-20479-4

[10] Gambhir KK, Bradley CJ. Characteristics of human erythrocyte insulin receptors. Diabetes. 1978;**27**(7):701-708. DOI: 10.2337/diab.27.7.701

[11] Zancan P, Sola-Penna M. Regulation of human erythrocyte metabolism by insulin: Cellular distribution of 6-phosphofructo-1-kinase and its implication for red blood cell function. Molecular Genetics and Metabolism. 2005;**86**(3):401-411. DOI: 10.1016/j.ymgme.2005.06.011

[12] Ferreira A, Rivera A, Romero JR. Na^+/Mg^{2+} exchange is functionally coupled to the insulin receptor. Journal of Cellular Physiology. 2004;**199**(3):434-440. DOI: 10.1002/jcp.10463

[13] Richards JP, Stephenson AH, Ellsworth ML, Sprague RS. Synergistic effects of C-peptide and insulin on low O_2-induced ATP release from human erythrocytes. American Journal of Physiology-Regulatory, Integrative and Comparative Physiology. 2013;**305**(11):R1331-R1336. DOI: 10.1152/ajpregu.00341.2013

[14] Hanson MS, Ellsworth ML, Achilleus D, Stephenson AH, Bowles EA,

Sridharan M, et al. Insulin inhibits low oxygen-induced ATP release from human erythrocytes: Implication for vascular control. Microcirculation. 2009;**16**(5):424-433. DOI: 10.1080/10739680902855218

[15] Dergunov SA, Bowles EA, Gordon W, Green M, Bierman A, Ellsworth ML, et al. Liposomal delivery of a phosphodiesterase 3 inhibitor rescues low oxygen-induced ATP release from erythrocytes of humans with type 2 diabetes. Biochemistry and Biophysics Reports. 2015;**2**:137-142. DOI: 10.1016/j.bbrep.2015.05.011

[16] Dubó S, Gallegos D, Cabrera L, Sobrevia L, Zúñiga L, González M. Cardiovascular action of insulin in health and disease: Endothelial L-arginine transport and cardiac voltage-dependent potassium channels. Frontiers in Physiology. 2016;**7**:74. DOI: 10.3389/fphys.2016.00074

[17] Wang Y, Yang P, Yan Z, Liu Z, Ma Q, Zhang Z, et al. The relationship between erythrocytes and diabetes mellitus. Journal of Diabetes Research. 2021;**2021**:1-9. DOI: 10.1155/2021/6656062

[18] Gupta P, Verma N, Bhattacharya S, Mahdi AA, Usman K, Tiwari S, et al. Association of Diabetic Autonomic Neuropathy with red blood cell aldose reductase activity. Canadian Journal of Diabetes. 2014;**38**(1):22-25. DOI: 10.1016/j.jcjd.2013.07.021

[19] Samaja M, Melotti D, Carenini A, Pozza G. Glycosylated haemoglobins and the oxygen affinity of whole blood. Diabetologia. 1982;**23**(5):399-402. DOI: 10.1007/BF00260950

[20] Sada K, Nishikawa T, Kukidome D, Yoshinaga T, Kajihara N, Sonoda K, et al. Hyperglycemia induces cellular hypoxia through production of mitochondrial ROS followed by suppression of Aquaporin-1. PLoS One. 2016;**11**(7):e0158619. DOI: 10.1371/journal.pone.0158619

[21] Polekhina G, Board PG, Gali RR, Rossjohn J, Parker MW. Molecular basis of glutathione synthetase deficiency and a rare gene permutation event. The EMBO Journal. 1999;**18**(12):3204-3213. DOI: 10.1093/emboj/18.12.3204

[22] Kumar P, Liu C, Hsu JW, Chacko S, Minard C, Jahoor F, et al. Glycine and N-acetylcysteine (GlyNAC) supplementation in older adults improves glutathione deficiency, oxidative stress, mitochondrial dysfunction, inflammation, insulin resistance, endothelial dysfunction, genotoxicity, muscle strength, and cognition: Results of a pilot clinical trial. Clinical and Translational Medicine. 2021;**11**(3):e372. DOI: 10.1002/ctm2.372

[23] Okamura T, Okada M, Kikuchi T, Wakizaka H, Zhang M-R. Mechanisms of glutathione-conjugate efflux from the brain into blood: Involvement of multiple transporters in the course. Journal of Cerebral Blood Flow and Metabolism. 2020;**40**(1):116-125. DOI: 10.1177/0271678X18808399

[24] Barrand MA, Winterberg M, Ng F, Nguyen M, Kirk K, Hladky SB. Glutathione export from human erythrocytes and *plasmodium falciparum* malaria parasites. Biochemical Journal. 2012;**448**(3):389-400. DOI: 10.1042/BJ20121050

[25] D'Orazio M, Pacello F, Battistoni A. Extracellular glutathione decreases the ability of Burkholderia cenocepacia to penetrate into epithelial cells and to induce an inflammatory response. PLoS One. 2012;7(10):e47550. DOI: 10.1371/journal.pone.0047550

[26] Helbling B, Von Overbeck J, Lauterburg BH. Decreased release of glutathione into the systemic circulation of patients with HIV infection. European Journal of Clinical Investigation. 1996;**26**(1):38-44. DOI: 10.1046/j.1365-2362.1996.88237.x

[27] Cao R, Kolloli A, Kumar R, Owens J, Sasaninia K, Vaughn C, et al. Effects of glutathione diminishment on the immune responses against mycobacterium tuberculosis infection. Applied Sciences. 2021;**11**(17):8274. DOI: 10.3390/app11178274

[28] Zhang Z, Zhang X, Fang X, Niimi M, Huang Y, Piao H, et al. Glutathione inhibits antibody and complement-mediated immunologic cell injury via multiple mechanisms. Redox Biology. 2017;**12**:571-581. DOI: 10.1016/j.redox.2017.03.030

[29] Ballatori N, Krance SM, Marchan R, Hammond CL. Plasma membrane glutathione transporters and their roles in cell physiology and pathophysiology. Molecular Aspects of Medicine. 2009;**30**(1-2):13-28. DOI: 10.1016/j.mam.2008.08.004

[30] Tukijan F, Chandrakanthan M, Nguyen LN. The signalling roles of sphingosine-1-phosphate derived from red blood cells and platelets: Circulatory sphingosine 1-phosphate. British Journal of Pharmacology. 2018;**175**(19):3741-3746. DOI: 10.1111/bph.14451

[31] Gazit SL, Mariko B, Thérond P, Decouture B, Xiong Y, Couty L, et al. Platelet and erythrocyte sources of S1P are redundant for vascular development and homeostasis, but both rendered essential after plasma S1P depletion in anaphylactic shock. Circulation Research. 2016;**119**(8). DOI: 10.1161/CIRCRESAHA.116.308929

[32] Wang Z, Min X, Xiao S-H, Johnstone S, Romanow W, Meininger D, et al. Molecular basis of sphingosine kinase 1 substrate recognition and catalysis. Structure. 2013;**21**(5):798-809. DOI: 10.1016/j.str.2013.02.025

[33] Nguyen TQ, Vu TM, Tukijan F, Muralidharan S, Foo JC, Li Chin JF, et al. Erythrocytes efficiently utilize exogenous sphingosines for S1P synthesis and export via Mfsd2b. Journal of Biological Chemistry. 2021;**296**:100201. DOI: 10.1074/jbc.RA120.012941

[34] Ito K, Anada Y, Tani M, Ikeda M, Sano T, Kihara A, et al. Lack of sphingosine 1-phosphate-degrading enzymes in erythrocytes. Biochemical and Biophysical Research Communications. 2007;**357**(1):212-217. DOI: 10.1016/j.bbrc.2007.03.123

[35] Christensen PM, Bosteen MH, Hajny S, Nielsen LB, Christoffersen C. Apolipoprotein M mediates sphingosine-1-phosphate efflux from erythrocytes. Scientific Reports. 2017;**7**(1):14983. DOI: 10.1038/s41598-017-15043-y

[36] Winkler MS, Claus RA, Schilder M, Pöhlmann S, Coldewey SM, Grundmann J, et al. Erythrocytes increase endogenous sphingosine 1-phosphate levels as an adaptive response to SARS-CoV-2 infection. Clinical Science. 2021;**135**(24):2781-2791. DOI: 10.1042/CS20210666

[37] Xie T, Chen C, Peng Z, Brown BC, Reisz JA, Xu P, et al. Erythrocyte metabolic reprogramming by sphingosine 1-phosphate in chronic kidney disease and therapies. Circulation Research. 2020;**127**(3):360-375. DOI: 10.1161/CIRCRESAHA.119.316298

[38] Nirmalraj PN, Schneider T, Felbecker A. Spatial organization of protein aggregates on red blood cells

as physical biomarkers of Alzheimer's disease pathology. Science Advances. 2021;**7**(39):eabj2137. DOI: 10.1126/sciadv.abj2137

[39] Mill J, Patel V, Okonkwo O, Li L, Raife T. Erythrocyte sphingolipid species as biomarkers of Alzheimer's disease. Journal of Pharmaceutical Analysis. 2022;**12**(1):178-185. DOI: 10.1016/j.jpha.2021.07.005

[40] Leeuw FA, Peeters CFW, Kester MI, Harms AC, Struys EA, Hankemeier T, et al. Blood-based metabolic signatures in Alzheimer's disease. Alzheimer's Dement. 2017;**8**(1):196-207. DOI: 10.1016/j.dadm.2017.07.006

[41] Winchester LM, Powell J, Lovestone S, Nevado-Holgado AJ. Red blood cell indices and anaemia as causative factors for cognitive function deficits and for Alzheimer's disease. Genome Medicine. 2018;**10**(1):51. DOI: 10.1186/s13073-018-0556-z

[42] Tenner AJ. Complement-mediated events in Alzheimer's disease: Mechanisms and potential therapeutic targets. Journal of Immunology. 2020;**204**(2):306-315. DOI: 10.4049/jimmunol.1901068

[43] Zhang D-F, Li J, Wu H, Cui Y, Bi R, Zhou H-J, et al. CFH variants affect structural and functional brain changes and genetic risk of Alzheimer's disease. Neuropsychopharmacology. 2016;**41**(4):1034-1045. DOI: 10.1038/npp.2015.232

[44] Torvell M, Carpanini SM, Daskoulidou N, Byrne RAJ, Sims R, Morgan BP. Genetic insights into the impact of complement in Alzheimer's disease. Genes. 2021;**12**(12):1990. DOI: 10.3390/genes12121990

[45] Johansson JU, Brubaker WD, Javitz H, Bergen AW, Nishita D, Trigunaite A, et al. Peripheral complement interactions with amyloid β peptide in Alzheimer's disease: Polymorphisms, structure, and function of complement receptor 1. Alzheimer's Dementia. 2018;**14**(11):1438-1449. DOI: 10.1016/j.jalz.2018.04.003

[46] Brubaker WD, Crane A, Johansson JU, Yen K, Garfinkel K, Mastroeni D, et al. Peripheral complement interactions with amyloid β peptide: Erythrocyte clearance mechanisms. Alzheimer's Dementia. 2017;**13**(12):1397-1409. DOI: 10.1016/j.jalz.2017.03.010

[47] Vaz M, Silva V, Monteiro C, Silvestre S. Role of aducanumab in the treatment of Alzheimer's disease: Challenges and opportunities. Clinical Interventions in Aging. 2022;**17**:797-810. DOI: 10.2147/CIA.S325026

[48] Crane A, Brubaker WD, Johansson JU, Trigunaite A, Ceballos J, Bradt B, et al. Peripheral complement interactions with amyloid β peptide in Alzheimer's disease: 2. Relationship to amyloid β immunotherapy. Alzheimer's. Dementia. 2018;**14**(2):243-252. DOI: 10.1016/j.jalz.2017.04.015

[49] Chen Y, Chu JMT, Chang RCC, Wong GTC. The complement system in the central nervous system: From neurodevelopment to neurodegeneration. Biomolecules. 2022;**12**(2):337. DOI: 10.3390/biom12020337

[50] Gomez-Arboledas A, Acharya MM, Tenner AJ. The role of complement in synaptic pruning and neurodegeneration. ImmunoTargets and Therapy. 2021;**10**:373-386. DOI: 10.2147/ITT.S305420

[51] Zhang L, Chen JY, Kerr C, Cobb BA, Maciejewski JP, Lin F. Reduced red blood cell surface level of factor H as a mechanism underlying paroxysmal

nocturnal hemoglobinuria. Leukemia. 2021;**35**(4):1176-1187. DOI: 10.1038/s41375-020-1008-5

[52] Mahmoudi R, Feldman S, Kisserli A, Duret V, Tabary T, Bertholon L-A, et al. Inherited and acquired decrease in complement receptor 1 (CR1) density on red blood cells associated with high levels of soluble CR1 in Alzheimer's disease. IJMS. 2018;**19**(8):2175. DOI: 10.3390/ijms19082175

[53] Antonelou MH, Kriebardis AG, Stamoulis KE, Trougakos IP, Papassideri IS. Apolipoprotein J/Clusterin in human erythrocytes is involved in the molecular process of defected material disposal during vesiculation. PLoS One. 2011;**6**(10):e26033. DOI: 10.1371/journal.pone.0026033

[54] Aikawa T, Holm M-L, Kanekiyo T. ABCA7 and pathogenic pathways of Alzheimer's disease. Brain Sciences. 2018;**8**(2):27. DOI: 10.3390/brainsci8020027

[55] Ammann EM, Pottala JV, Robinson JG, Espeland MA, Harris WS. Erythrocyte omega-3 fatty acids are inversely associated with incident dementia: Secondary analyses of longitudinal data from the Women's health initiative memory study (WHIMS). Prostaglandins, Leukotrienes and Essential Fatty Acids. 2017;**121**:68-75. DOI: 10.1016/j.plefa.2017.06.006

[56] Carelli-Alinovi C, Misiti F. Erythrocytes as potential link between diabetes and Alzheimer's disease. Frontiers in Aging Neuroscience. 2017;**9**:276. DOI: 10.3389/fnagi.2017.00276

[57] Várady G, Szabó E, Fehér Á, Németh A, Zámbó B, Pákáski M, et al. Alterations of membrane protein expression in red blood cells of Alzheimer's disease patients. Alzheimer's Dementia. 2015;**1**(3):334-338. DOI: 10.1016/j.dadm.2015.06.007

[58] Kosenko EA, Tikhonova LA, Montoliu C, Barreto GE, Aliev G, Kaminsky YG. Metabolic abnormalities of erythrocytes as a risk factor for Alzheimer's disease. Frontiers in Neuroscience. 2018;**11**:728. DOI: 10.3389/fnins.2017.00728

[59] Xiong W, Liang Y, Li X, Liu G, Wang Z. A direct quantitative analysis of erythrocyte intracellular ionized magnesium in physiological and pathological conditions. Biological & Pharmaceutical Bulletin. 2019;**42**(3):357-364. DOI: 10.1248/bpb.b18-00406

[60] Sitzia C, Sterlicchio M, Crapanzano C, Dozio E, Vianello E, Corsi Romanelli MM. Intra-erythrocytes magnesium deficiency could reflect cognitive impairment status due to vascular disease: A pilot study. Journal of Translational Medicine. 2020;**18**(1):458. DOI: 10.1186/s12967-020-02645-w

[61] Iadecola C. The pathobiology of vascular dementia. Neuron. 2013;**80**(4):844-866. DOI: 10.1016/j.neuron.2013.10.008

[62] Mossanen Parsi M, Duval C, Ariëns RAS. Vascular dementia and crosstalk between the complement and coagulation systems. Front Cardiovasc Med. 2021;**8**:803169. DOI: 10.3389/fcvm.2021.803169

[63] Loures C d MG, RCF D, MVF S, Cicarini WB, de Souza LC, Caramelli P, et al. Hemostatic abnormalities in dementia: A systematic review and Meta-analysis. Seminars in Thrombosis and Hemostasis. 2019;**45**(05):514-522. DOI: 10.1055/s-0039-1688444

[64] Liou AY, Molday LL, Wang J, Andersen JP, Molday RS. Identification and functional analyses of disease-associated P4-ATPase phospholipid flippase variants in red blood cells. Journal of Biological Chemistry. 2019;**294**(17):6809-6821. DOI: 10.1074/jbc.RA118.007270

[65] Nakanishi H, Irie K, Segawa K, Hasegawa K, Fujiyoshi Y, Nagata S, et al. Crystal structure of a human plasma membrane phospholipid flippase. Journal of Biological Chemistry. 2020;**295**(30):10180-10194. DOI: 10.1074/jbc.RA120.014144

[66] Timcenko M, Dieudonné T, Montigny C, Boesen T, Lyons JA, Lenoir G, et al. Structural basis of substrate-independent phosphorylation in a P4-ATPase lipid Flippase. Journal of Molecular Biology. 2021;**433**(16):167062. DOI: 10.1016/j.jmb.2021.167062

[67] Dieudonné T, Herrera SA, Laursen MJ, Lejeune M, Stock C, Slimani K, et al. Autoinhibition and regulation by phosphoinositides of ATP8B1, a human lipid flippase associated with intrahepatic cholestatic disorders. eLife. 2022;**11**:e75272. DOI: 10.7554/eLife.75272

[68] Seki M, Arashiki N, Takakuwa Y, Nitta K, Nakamura F. Reduction in flippase activity contributes to surface presentation of phosphatidylserine in human senescent erythrocytes. Journal of Cellular and Molecular Medicine. 2020;**24**(23):13991-14000. DOI: 10.1111/jcmm.16010

[69] Arashiki N, Takakuwa Y, Mohandas N, Hale J, Yoshida K, Ogura H, et al. ATP11C is a major flippase in human erythrocytes and its defect causes congenital hemolytic anemia. Haematologica. 2016;**101**(5):559-565. DOI: 10.3324/haematol.2016.142273

[70] Kodigepalli KM, Bowers K, Sharp A, Nanjundan M. Roles and regulation of phospholipid scramblases. FEBS Letters. 2015;**589**(1):3-14. DOI: 10.1016/j.febslet.2014.11.036

[71] Öhlinger T, Müllner EW, Fritz M, Sauer T, Werning M, Baron DM, et al. Lysophosphatidic acid-induced pro-thrombotic phosphatidylserine exposure and ionophore-induced microvesiculation is mediated by the scramblase TMEM16F in erythrocytes. Blood Cells, Molecules, and Diseases. 2020;**83**:102426. DOI: 10.1016/j.bcmd.2020.102426

[72] Yan M, Xu M, Li Z, An Y, Wang Z, Li S, et al. TMEM16F mediated phosphatidylserine exposure and microparticle release on erythrocyte contribute to hypercoagulable state in hyperuricemia. Blood Cells, Molecules, and Diseases. 2022;**96**:102666. DOI: 10.1016/j.bcmd.2022.102666

[73] Alvadia C, Lim NK, Mosina VC, Oostergetel GT, Dutzler R, Paulino C. Cryo-EM structures and functional characterization of the murine lipid scramblase TMEM16F. eLife. 2019;**8**:e44365. DOI: 10.7554/eLife.44365.001

[74] Pomorski TG, Menon AK. Lipid somersaults: Uncovering the mechanisms of protein-mediated lipid flipping. Progress in Lipid Research. 2016;**64**:69-84. DOI: 10.1016/j.plipres.2016.08.003

[75] Suzuki J, Umeda M, Sims PJ, Nagata S. Calcium-dependent phospholipid scrambling by TMEM16F. Nature. 2010;**468**(7325):834-838. DOI: 10.1038/nature09583

[76] Castoldi E, Collins PW, Williamson PL, Bevers EM. Compound heterozygosity for 2 novel

TMEM16F mutations in a patient with Scott syndrome. Blood. 2011;**117**(16):4399-4400. DOI: 10.1182/blood-2011-01-332502

[77] Bevers E, Wiedmer T, Comfurius P, Shattil S, Weiss H, Zwaal R, et al. Defective Ca(2+)-induced microvesiculation and deficient expression of procoagulant activity in erythrocytes from a patient with a bleeding disorder: A study of the red blood cells of Scott syndrome. Blood. 1992;**79**(2):380-388. DOI: 10.1182/blood.V79.2.380.380

[78] Lin H, Roh J, Woo JH, Kim SJ, Nam JH. TMEM16F/ANO6, a Ca^{2+}-activated anion channel, is negatively regulated by the actin cytoskeleton and intracellular MgATP. Biochemical and Biophysical Research Communications. 2018;**503**(4):2348-2354. DOI: 10.1016/j.bbrc.2018.06.160

[79] Grabarek Z. Insights into modulation of calcium signaling by magnesium in calmodulin, troponin C and related EF-hand proteins. Biochimica et Biophysica Acta (BBA)—Molecular Cell Research. 2011;**1813**(5):913-921. DOI: 10.1016/j.bbamcr.2011.01.017

[80] Talbi K, Ousingsawat J, Centeio R, Schreiber R, Kunzelmann K. Calmodulin-dependent regulation of overexpressed but not endogenous TMEM16A expressed in airway epithelial cells. Membranes. 2021;**11**(9):723. DOI: 10.3390/membranes11090723

[81] Zhang D-L, Wu J, Shah BN, Greutélaers KC, Ghosh MC, Ollivierre H, et al. Erythrocytic ferroportin reduces intracellular iron accumulation, hemolysis, and malaria risk. Science. 2018;**359**(6383):1520-1523. DOI: 10.1126/science.aal2022

[82] Zhang D-L, Ghosh MC, Ollivierre H, Li Y, Rouault TA. Ferroportin deficiency in erythroid cells causes serum iron deficiency and promotes hemolysis due to oxidative stress. Blood. 2018;**132**(19):2078-2087. DOI: 10.1182/blood-2018-04-842997

[83] Ganz T. Erythrocytes and erythroblasts give up iron. Blood. 2018;**132**(19):2004-2005. DOI: 10.1182/blood-2018-09-876102

[84] Stockwell BR. Ferroptosis turns 10: Emerging mechanisms, physiological functions, and therapeutic applications. Cell. 2022;**185**(14):2401-2421. DOI: 10.1016/j.cell.2022.06.003

[85] Sun Y, Xia X, Basnet D, Zheng JC, Huang J, Liu J. Mechanisms of ferroptosis and emerging links to the pathology of neurodegenerative diseases. Frontiers in Aging Neuroscience. 2022;**14**:904152. DOI: 10.3389/fnagi.2022.904152

[86] Majerníková N, den Dunnen WFA, Dolga AM. The potential of ferroptosis-targeting therapies for Alzheimer's disease: From mechanism to transcriptomic analysis. Frontiers in Aging Neuroscience. 2021;**13**:745046. DOI: 10.3389/fnagi.2021.745046

[87] Yan N, Zhang J. Iron metabolism, ferroptosis, and the links with Alzheimer's disease. Frontiers in Neuroscience. 2020;**13**:1443. DOI: 10.3389/fnins.2019.01443

[88] Carelli-Alinovi C, Dinarelli S, Sampaolese B, Misiti F, Girasole M. Morphological changes induced in erythrocyte by amyloid beta peptide and glucose depletion: A combined atomic force microscopy and biochemical study. Biochimica et Biophysica Acta (BBA)—Biomembranes. 2019;**1861**(1):236-244. DOI: 10.1016/j.bbamem.2018.07.009

[89] Dinarelli S, Girasole M, Misiti F. Amyloid β peptide affects erythrocyte morphology: Role of intracellular

signaling pathways. Clinical Hemorheology and Microcirculation. 2019;**71**(4):437-449. DOI: 10.3233/CH-199007

[90] Misiti F, Orsini F, Clementi ME, Masala D, Tellone E, Galtieri A, et al. Amyloid peptide inhibits ATP release from human erythrocytes. Biochemistry and Cell Biology. 2008;**86**(6):501-508. DOI: 10.1139/O08-139

[91] Pretorius E. The adaptability of red blood cells. Cardiovascular Diabetology. 2013;**12**(1):63. DOI: 10.1186/1475-2840-12-63

[92] Khan AA, Hadi Y, Hassan A, Kupec J. Polycythemia and Anemia in hereditary hemochromatosis. Cureus. 2020;**12**(4):e7607. DOI: 10.7759/cureus.7607

[93] Aslam M, Gündüz D, Troidl C, Heger J, Hamm CW, Schulz R. Purinergic regulation of endothelial barrier function. IJMS. 2021;**22**(3):1207. DOI: 10.3390/ijms22031207

[94] Zhou J, Zhou Z, Liu X, Yin H-Y, Tang Y, Cao X. P2X7 receptor-mediated inflammation in cardiovascular disease. Frontiers in Pharmacology. 2021;**12**:654425. DOI: 10.3389/fphar.2021.654425

[95] Jiang L-H, Caseley EA, Muench SP, Roger S. Structural basis for the functional properties of the P2X7 receptor for extracellular ATP. Purinergic Signalling. 2021;**17**(3):331-344. DOI: 10.1007/s11302-021-09790-x

[96] McCarthy AE, Yoshioka C, Mansoor SE. Full-length P2X7 structures reveal how palmitoylation prevents channel desensitization. Cell. 2019;**179**(3):659-670.e13. DOI: 10.1016/j.cell.2019.09.017

[97] He X, Zhang Y, Xu Y, Xie L, Yu Z, Zheng J. Function of the P2X7 receptor in hematopoiesis and leukemogenesis. Experimental Hematology. 2021;**104**:40-47. DOI: 10.1016/j.exphem.2021.10.001

[98] Patritti-Cram J, Coover RA, Jankowski MP, Ratner N. Purinergic signaling in peripheral nervous system glial cells. Glia. 2021;**69**(8):1837-1851. DOI: 10.1002/glia.23969

[99] Hechler B, Gachet C. Purinergic receptors in thrombosis and inflammation. ATVB. 2015;**35**(11):2307-2315. DOI: 10.1161/ATVBAHA.115.303395

[100] Ozen M, Xie H, Shin N, Al Yousif G, Clemens J, McLane MW, et al. Magnesium sulfate inhibits inflammation through P2X7 receptors in human umbilical vein endothelial cells. Pediatric Research. 2020;**87**(3):463-471. DOI: 10.1038/s41390-019-0557-7

[101] Zhou Z. Purinergic interplay between erythrocytes and platelets in diabetes-associated vascular dysfunction. Purinergic Signalling. 2021;**17**(4):705-712. DOI: 10.1007/s11302-021-09807-5

[102] Zimmermann H. Ectonucleoside triphosphate diphosphohydrolases and ecto-5′-nucleotidase in purinergic signaling: How the field developed and where we are now. Purinergic Signalling. 2021;**17**(1):117-125. DOI: 10.1007/s11302-020-09755-6

[103] Minor M, Alcedo KP, Battaglia RA, Snider NT. Cell type- and tissue-specific functions of ecto-5′-nucleotidase (CD73). American Journal of Physiology-Cell Physiology. 2019;**317**(6):C1079-C1092. DOI: 10.1152/ajpcell.00285.2019

[104] Vivian JP, Riedmaier P, Ge H, Le Nours J, Sansom FM, Wilce MCJ,

et al. Crystal structure of a legionella pneumophila ecto-triphosphate diphosphohydrolase, a structural and functional homolog of the eukaryotic NTPDases. Structure. 2010;**18**(2):228-238. DOI: 10.1016/j.str.2009.11.014

[105] Knapp K, Zebisch M, Pippel J, El-Tayeb A, Müller CE, Sträter N. Crystal structure of the human ecto-5′-nucleotidase (CD73): Insights into the regulation of purinergic signaling. Structure. 2012;**20**(12):2161-2173. DOI: 10.1016/j.str.2012.10.001

[106] Lee GR, Shaefi S, Otterbein LE. HO-1 and CD39: It takes two to protect the realm. Frontiers in Immunology. 2019;**10**:1765. DOI: 10.3389/fimmu.2019.01765

[107] Hawkins CF, Bagnara AS. Adenosine kinase from human erythrocytes: Kinetic studies and characterization of adenosine binding sites. Biochemistry. 1987;**26**(7):1982-1987. DOI: 10.1021/bi00381a030

[108] Boison D, Jarvis MF. Adenosine kinase: A key regulator of purinergic physiology. Biochemical Pharmacology. 2021;**187**:114321. DOI: 10.1016/j.bcp.2020.114321

[109] Song Y, Tang L, Han J, Gao Y, Tang B, Shao M, et al. Uric acid provides protective role in red blood cells by antioxidant defense: A hypothetical analysis. Oxidative Medicine and Cellular Longevity. 2019;**2019**:1-12. DOI: 10.1155/2019/3435174

[110] Luo M, Li Z-Z, Li Y-Y, Chen L-Z, Yan S-P, Chen P, et al. Relationship between red cell distribution width and serum uric acid in patients with untreated essential hypertension. Scientific Reports. 2015;**4**(1):7291. DOI: 10.1038/srep07291

[111] Hershfield MS, Roberts LJ, Ganson NJ, Kelly SJ, Santisteban I, Scarlett E, et al. Treating gout with pegloticase, a PEGylated urate oxidase, provides insight into the importance of uric acid as an antioxidant in vivo. Proceedings of the National Academy of Sciences of the United States of America. 2010;**107**(32):14351-14356. DOI: 10.1073/pnas.1001072107

[112] Karsten E, Breen E, Herbert BR. Red blood cells are dynamic reservoirs of cytokines. Scientific Reports. 2018;**8**(1):3101. DOI: 10.1038/s41598-018-21387-w

[113] Ayi K, Lu Z, Serghides L, Ho JM, Finney C, Wang JCY, et al. CD47-SIRPα interactions regulate macrophage uptake of plasmodium falciparum-infected erythrocytes and clearance of malaria in vivo. Infection and Immunity. 2016;**84**(7):2002-2011. DOI: 10.1128/IAI.01426-15

[114] Li Z-Z, Zhang Z-W, Wang H, Chen Y-Q, Zhou X-F, Duan L-S, et al. Autologous transfusion of "old" red blood cells-induced M2 macrophage polarization through IL-10-Nrf2-HO-1 signaling complexes. Advances in Clinical and Experimental Medicine. 2020;**29**(7):833-840. DOI: 10.17219/acem/121924

[115] Klei TRL, Meinderts SM, van den Berg TK, van Bruggen R. From the cradle to the grave: The role of macrophages in erythropoiesis and erythrophagocytosis. Frontiers in Immunology. 2017;**8**:73. DOI: 10.3389/fimmu.2017.00073

[116] Li W, Guo R, Song Y, Jiang Z. Erythroblastic Island macrophages shape normal erythropoiesis and drive associated disorders in erythroid hematopoietic diseases. Frontiers in Cell and Development Biology. 2021;**8**:613885. DOI: 10.3389/fcell.2020.613885

[117] Liao C, Hardison RC, Kennett MJ, Carlson BA, Paulson RF, Prabhu KS. Selenoproteins regulate stress erythroid progenitors and spleen microenvironment during stress erythropoiesis. Blood. 2018;**131**(23):2568-2580. DOI: 10.1182/blood-2017-08-800607

[118] Kawatani Y, Suzuki T, Shimizu R, Kelly VP, Yamamoto M. Nrf2 and selenoproteins are essential for maintaining oxidative homeostasis in erythrocytes and protecting against hemolytic anemia. Blood. 2011;**117**(3):986-996. DOI: 10.1182/blood-2010-05-285817

[119] Altamura S, Vegi NM, Hoppe PS, Schroeder T, Aichler M, Walch A, et al. Glutathione peroxidase 4 and vitamin E control reticulocyte maturation, stress erythropoiesis and iron homeostasis. Haematologica. 2020;**105**(4):937-950. DOI: 10.3324/haematol.2018.212977

[120] Huang L, Mao X, Li Y, Liu D, Fan K, Liu R, et al. Multiomics analyses reveal a critical role of selenium in controlling T cell differentiation in Crohn's disease. Immunity. 2021;**54**(8):1728-1744.e7. DOI: 10.1016/j.immuni.2021.07.004

[121] Lv H, Wei G-Y, Guo C-S, Deng Y-F, Jiang Y-M, Gao C, et al. 20S proteasome and glyoxalase 1 activities decrease in erythrocytes derived from Alzheimer's disease patients. Neural Regeneration Research. 2020;**15**(1):178. DOI: 10.4103/1673-5374.264473

[122] Saito Y, Akazawa-Ogawa Y, Matsumura A, Saigoh K, Itoh S, Sutou K, et al. Oxidation and interaction of DJ-1 with 20S proteasome in the erythrocytes of early stage Parkinson's disease patients. Scientific Reports. 2016;**6**(1):30793. DOI: 10.1038/srep30793

[123] Gerner MC, Bileck A, Janker L, Ziegler LS, Öhlinger T, Raeven P, et al. Packed red blood cells inhibit T-cell activation via ROS-dependent signaling pathways. Journal of Biological Chemistry. 2021;**296**:100487. DOI: 10.1016/j.jbc.2021.100487

[124] Fonseca AM, Porto G, Uchida K, Arosa FA. Red blood cells inhibit activation-induced cell death and oxidative stress in human peripheral blood T lymphocytes. Blood. 2001;**97**(10):3152-3160. DOI: 10.1182/blood.v97.10.3152

[125] Karsten E, Breen E, McCracken SA, Clarke S, Herbert BR. Red blood cells exposed to cancer cells in culture have altered cytokine profiles and immune function. Scientific Reports. 2020;**10**(1):7727. DOI: 10.1038/s41598-020-64319-3

[126] Kanda A, Adachi T, Kayaba H, Yamada Y, Ueki S, Yamaguchi K, et al. Red blood cells regulate eosinophil chemotaxis by scavenging RANTES secreted from endothelial cells. Clinical and Experimental Allergy. 2004;**34**(10):1621-1626. DOI: 10.1111/j.1365-2222.2004.02073.x

[127] Mahdi A, Cortese-Krott MM, Kelm M, Li N, Pernow J. Novel perspectives on redox signaling in red blood cells and platelets in cardiovascular disease. Free Radical Biology and Medicine. 2021;**168**:95-109. DOI: 10.1016/j.freeradbiomed.2021.03.020

[128] Tejero J, Shiva S, Gladwin MT. Sources of vascular nitric oxide and reactive oxygen species and their regulation. Physiological Reviews. 2019;**99**(1):311-379. DOI: 10.1152/physrev.00036.2017

[129] Lehners M, Dobrowinski H, Feil S, Feil R. cGMP signaling and vascular smooth muscle cell plasticity. JCDD. 2018;**5**(2):20. DOI: 10.3390/jcdd5020020

[130] Deng J-T, Bhaidani S, Sutherland C, MacDonald JA, Walsh MP. Rho-associated kinase and zipper-interacting protein kinase, but not myosin light chain kinase, are involved in the regulation of myosin phosphorylation in serum-stimulated human arterial smooth muscle cells. PLoS One. 2019;**14**(12):e0226406. DOI: 10.1371/journal.pone.0226406

[131] Kiss A, Erdődi F, Lontay B. Myosin phosphatase: Unexpected functions of a long-known enzyme. Biochimica et Biophysica Acta (BBA)—Molecular Cell Research. 2019;**1866**(1):2-15. DOI: 10.1016/j.bbamcr.2018.07.023

[132] Cortese-Krott MM, Mergia E, Kramer CM, Lückstädt W, Yang J, Wolff G, et al. Identification of a soluble guanylate cyclase in RBCs: Preserved activity in patients with coronary artery disease. Redox Biology. 2018;**14**:328-337. DOI: 10.1016/j.redox.2017.08.020

[133] Wang G-R, Zhu Y, Halushka PV, Lincoln TM, Mendelsohn ME. Mechanism of platelet inhibition by nitric oxide: In vivo phosphorylation of thromboxane receptor by cyclic GMP-dependent protein kinase. Proceedings of the National Academy of Sciences of the United States of America. 1998;**95**(9):4888-4893. DOI: 10.1073/pnas.95.9.4888

[134] Bibli S-I, Papapetropoulos A, Iliodromitis EK, Daiber A, Randriamboavonjy V, Steven S, et al. Nitroglycerine limits infarct size through S-nitrosation of cyclophilin D: A novel mechanism for an old drug. Cardiovascular Research. 2019;**115**(3):625-636. DOI: 10.1093/cvr/cvy222

[135] Kolesnik B, Heine CL, Schmidt R, Schmidt K, Mayer B, Gorren ACF. Aerobic nitric oxide-induced thiol nitrosation in the presence and absence of magnesium cations. Free Radical Biology and Medicine. 2014;**76**:286-298. DOI: 10.1016/j.freeradbiomed.2014.08.024

[136] Zhou Z, Mahdi A, Tratsiakovich Y, Zahorán S, Kövamees O, Nordin F, et al. Erythrocytes from patients with type 2 diabetes induce endothelial dysfunction via arginase I. Journal of the American College of Cardiology. 2018;**72**(7):769-780. DOI: 10.1016/j.jacc.2018.05.052

[137] Wischmann P, Kuhn V, Suvorava T, Muessig JM, Fischer JW, Isakson BE, et al. Anaemia is associated with severe RBC dysfunction and a reduced circulating NO pool: Vascular and cardiac eNOS are crucial for the adaptation to anaemia. Basic Research in Cardiology. 2020;**115**(4):43. DOI: 10.1007/s00395-020-0799-x

[138] Daiber A, Münzel T. Interplay of the red blood cell and vascular endothelial nitric oxide synthase system to combat cardiac complications of anemia. Basic Research in Cardiology. 2020;**115**(4):44. DOI: 10.1007/s00395-020-0801-7

[139] Yamaguchi T, Manaka C, Ogura A, Nagadome S. Importance of cholesterol side chain in the membrane stability of human erythrocytes. Biological & Pharmaceutical Bulletin. 2021;**44**(6):888-893

[140] Hayakawa EH, Yamaguchi K, Mori M, Nardone G. Real-time cholesterol sorting in plasmodium falciparum-erythrocytes as revealed by 3D label-free imaging. Scientific Reports. 2020;**10**(1):2794. DOI: 10.1038/s41598-020-59552-9

[141] Ahiya AI, Bhatnagar S, Morrisey JM, Beck JR, Vaidya AB. Dramatic consequences of reducing erythrocyte membrane cholesterol on

plasmodium falciparum. Microbiology Spectrum. 2022;**10**(1):e00158-e00122

[142] Lankinen M, Uusitupa M, Schwab U. Genes and dietary fatty acids in regulation of fatty acid composition of plasma and erythrocyte membranes. Nutrients. 2018;**10**(11):1785. DOI: 10.3390/nu10111785

[143] Balta I, Stef L, Pet I, Iancu T, Stef D, Corcionivoschi N. Essential fatty acids as biomedicines in cardiac health. Biomedicine. 2021;**9**(10):1466. DOI: 10.3390/biomedicines9101466

[144] Hwang DH, Kim J-A, Lee JY. Mechanisms for the activation of toll-like receptor 2/4 by saturated fatty acids and inhibition by docosahexaenoic acid. European Journal of Pharmacology. 2016;**785**:24-35. DOI: 10.1016/j.ejphar.2016.04.024

[145] Harris WS, Tintle NL, Etherton MR, Vasan RS. Erythrocyte long-chain omega-3 fatty acid levels are inversely associated with mortality and with incident cardiovascular disease: The Framingham heart study. Journal of Clinical Lipidology. 2018;**12**(3):718-727.e6. DOI: 10.1016/j.jacl.2018.02.010

[146] Harris W, Tintle N, Ramachandran V. Erythrocyte n-6 fatty acids and risk for cardiovascular outcomes and total mortality in the Framingham heart study. Nutrients. 2018;**10**(12):2012. DOI: 10.3390/nu10122012

[147] Chen Q, Wang J, Wang J, Lin J, Chen L, Lin L, et al. Erythrocyte ω-3 polyunsaturated fatty acids are inversely associated with the risk of oral cancer: A case-control study. Nutrition & Diabetes. 2020;**10**(1):35. DOI: 10.1038/s41387-020-00140-1

[148] Niinistö S, Erlund I, Lee H-S, Uusitalo U, Salminen I, Aronsson CA, et al. Children's erythrocyte fatty acids are associated with the risk of islet autoimmunity. Scientific Reports. 2021;**11**(1):3627. DOI: 10.1038/s41598-021-82200-9

[149] Goozee K, Chatterjee P, James I, Shen K, Sohrabi HR, Asih PR, et al. Alterations in erythrocyte fatty acid composition in preclinical Alzheimer's disease. Scientific Reports. 2017;**7**(1):676. DOI: 10.1038/s41598-017-00751-2

[150] Li N, Yang P, Tang M, Liu Y, Guo W, Lang B, et al. Reduced erythrocyte membrane polyunsaturated fatty acid levels indicate diminished treatment response in patients with multi- versus first-episode schizophrenia. Schizophrenia. 2022;**8**(1):7. DOI: 10.1038/s41537-022-00214-2

[151] Kim S-W, Jhon M, Kim J-M, Smesny S, Rice S, Berk M, et al. Relationship between erythrocyte fatty acid composition and psychopathology in the Vienna Omega-3 study. PLoS One. 2016;**11**(3):e0151417. DOI: 10.1371/journal.pone.0151417

[152] McNamara RK, Welge JA. Meta-analysis of erythrocyte polyunsaturated fatty acid biostatus in bipolar disorder. Bipolar Disorders. 2016;**18**(3):300-306. DOI: 10.1111/bdi.12386

[153] Howsmon DP, Adams JB, Kruger U, Geis E, Gehn E, Hahn J. Erythrocyte fatty acid profiles in children are not predictive of autism spectrum disorder status: A case control study. Biomarker Research. 2018;**6**(1):12. DOI: 10.1186/s40364-018-0125-z

[154] Sun Q, Ma J, Campos H, Hankinson SE, Manson JE, Stampfer MJ, et al. A prospective study of trans fatty acids in erythrocytes and risk of coronary heart disease. Circulation.

2007;**115**(14):1858-1865. DOI: 10.1161/CIRCULATIONAHA.106.679985

[155] Liu X-R, Deng Z-Y, Hu J-N, Fan Y-W, Liu R, Li J, et al. Erythrocyte membrane trans -fatty acid index is positively associated with a 10-year CHD risk probability. The British Journal of Nutrition. 2013;**109**(9):1695-1703. DOI: 10.1017/S0007114513000196

[156] von Schacky C, Passow A, Kiefl R. Trans-fatty acid levels in erythrocytes in Europe. European Journal of Nutrition. 2017;**56**(4):1719-1723. DOI: 10.1007/s00394-016-1218-9

[157] Thomas HB, Moots RJ, Edwards SW, Wright HL. Whose gene is it anyway? The effect of preparation purity on neutrophil transcriptome studies. PLoS One. 2015;**10**(9):e0138982. DOI: 10.1371/journal.pone.0138982

[158] Jain V, Yang W-H, Wu J, Roback JD, Gregory SG, Chi J-T. Single cell RNA-Seq analysis of human red cells. Frontiers in Physiology. 2022;**13**:828700. DOI: 10.3389/fphys.2022.828700

[159] Sun L, Yu Y, Niu B, Wang D. Red blood cells as potential repositories of MicroRNAs in the circulatory system. Frontiers in Genetics. 2020;**11**:442. DOI: 10.3389/fgene.2020.00442

[160] Song H, Fang X, Jin L, Shaw GX, Wang Y-X, Ji X. The functional cycle of Rnt1p: Five consecutive steps of double-stranded RNA processing by a eukaryotic RNase III. Structure. 2017;**25**(2):353-363. DOI: 10.1016/j.str.2016.12.013

[161] Dharavath S, Shaw GX, Ji X. Structural basis for dicer-like function of an engineered RNase III variant and insights into the reaction trajectory of two-Mg^{2+}-ion catalysis. RNA Biology. 2022;**19**(1):908-915. DOI: 10.1080/15476286.2022.2099650

[162] Nam S, Ryu H, Son W, Kim YH, Kim KT, Balch C, et al. Mg^{2+} effect on argonaute and RNA duplex by molecular dynamics and bioinformatics implications. PLoS One. 2014;**9**(10):e109745. DOI: 10.1371/journal.pone.0109745

[163] Nath A, Rayabaram J, Ijee S, Bagchi A, Chaudhury AD, Roy D, et al. Comprehensive analysis of microRNAs in human adult erythropoiesis. Cell. 2021;**10**(11):3018. DOI: 10.3390/cells10113018

[164] Cloos A-S, Ghodsi M, Stommen A, Vanderroost J, Dauguet N, Pollet H, et al. Interplay between plasma membrane lipid alteration, oxidative stress and calcium-based mechanism for extracellular vesicle biogenesis from erythrocytes during blood storage. Frontiers in Physiology. 2020;**11**:712. DOI: 10.3389/fphys.2020.00712

[165] Koga Y, Yasunaga M, Moriya Y, Akasu T, Fujita S, Yamamoto S. Exosome can prevent RNase from degrading microRNA in feces. Journal of Gastrointestinal Oncology. 2011;**2**(4):8. DOI: 10.3978/j.issn.2078-6891.2011.015

[166] Wang T, Wu F, Yu D. miR-144/451 in hematopoiesis and beyond. ExRNA. 2019;**1**(1):16. DOI: 10.1186/s41544-019-0035-8

[167] Mantel P-Y, Hjelmqvist D, Walch M, Kharoubi-Hess S, Nilsson S, Ravel D, et al. Infected erythrocyte-derived extracellular vesicles alter vascular function via regulatory Ago2-miRNA complexes in malaria. Nature Communications. 2016;**7**(1):12727. DOI: 10.1038/ncomms12727

[168] Chamnanchanunt S, Kuroki C, Desakorn V, Enomoto M, Thanachartwet V, Sahassananda D, et al. Downregulation of plasma miR-451 and

miR-16 in plasmodium vivax infection. Experimental Parasitology. 2015;**155**:19-25. DOI: 10.1016/j.exppara.2015.04.013

[169] Ketprasit N, Cheng IS, Deutsch F, Tran N, Imwong M, Combes V, et al. The characterization of extracellular vesicles-derived microRNAs in Thai malaria patients. Malaria Journal. 2020;**19**(1):285. DOI: 10.1186/s12936-020-03360-z

[170] Atamna H, Ginsburg H. The malaria parasite supplies glutathione to its host cell—investigation of glutathione transport and metabolism in human erythrocytes infected with plasmodium falciparum. European Journal of Biochemistry. 1997;**250**(3):670-679. DOI: 10.1111/j.1432-1033.1997.00670.x

[171] Mbagwu SI, Lannes N, Walch M, Filgueira L, Mantel P-Y. Human microglia respond to malaria-induced extracellular vesicles. Pathogens. 2019;**9**(1):21. DOI: 10.3390/pathogens9010021

[172] Hernández-Castañeda MA, Lavergne M, Casanova P, Nydegger B, Merten C, Subramanian BY, et al. A profound membrane reorganization defines susceptibility of plasmodium falciparum infected red blood cells to lysis by Granulysin and perforin. Frontiers in Immunology. 2021;**12**:643746. DOI: 10.3389/fimmu.2021.643746

[173] Albrecht-Schgoer K, Lackner P, Schmutzhard E, Baier G. Cerebral malaria: Current clinical and immunological aspects. Frontiers in Immunology. 2022;**13**:863568. DOI: 10.3389/fimmu.2022.863568

[174] Sánchez-Sánchez R, Gómez-Ferrer M, Reinal I, Buigues M, Villanueva-Bádenas E, Ontoria-Oviedo I, et al. miR-4732-3p in extracellular vesicles from mesenchymal stromal cells is cardioprotective during myocardial ischemia. Frontiers in Cell and Development Biology. 2021;**9**:734143. DOI: 10.3389/fcell.2021.734143

[175] Hasse S, Duchez A-C, Fortin P, Boilard E, Bourgoin SG. Interplay between LPA2 and LPA3 in LPA-mediated phosphatidylserine cell surface exposure and extracellular vesicles release by erythrocytes. Biochemical Pharmacology. 2021;**192**:114667. DOI: 10.1016/j.bcp.2021.114667

[176] Oleinikov AV. Malaria parasite plasmodium falciparum proteins on the surface of infected erythrocytes as targets for novel drug discovery. Biochemistry Moscow. 2022;**87**(S1):S192-S202. DOI: 10.1134/S0006297922140152

[177] Ulfberg J, Stehlik R. Finland's handling of selenium is a model in these times of coronavirus infections. The British Journal of Nutrition. 2021;**125**(12):1437-1438. DOI: 10.1017/S0007114520003827

[178] Giacosa A, Faliva M, Perna S, Minoia C, Ronchi A, Rondanelli M. Selenium fortification of an Italian rice cultivar via foliar fertilization with sodium selenate and its effects on human serum selenium levels and on erythrocyte glutathione peroxidase activity. Nutrients. 2014;**6**(3):1251-1261. DOI: 10.3390/nu6031251

[179] Achouba A, Dumas P, Ouellet N, Little M, Lemire M, Ayotte P. Selenoneine is a major selenium species in beluga skin and red blood cells of Inuit from Nunavik. Chemosphere. 2019;**229**:549-558. DOI: 10.1016/j.chemosphere.2019.04.191

Chapter 3

Hormones Action on Erythrocytes and Signaling Pathways

Camila Cristina Guimarães-Nobre, Evelyn Mendonça-Reis, Lyzes Rosa Teixeira-Alves and Clemilson Berto Junior

Abstract

Erythrocytes are the most abundant cell type in the human body, although considered as merely hemoglobin carriers for a long time. Extensive studies on its biochemical pathways, metabolism, and structure-activity relationship with a consistent number of publications demonstrated the presence of autocrine, paracrine, and endocrine hormone receptors. In this chapter, some of these hormones will be discussed, bringing attention to those that regulate erythrocyte survival, disease connection, and functionality.

Keywords: hormones, signaling pathway, TSH, endothelim-1, angiotensin II

1. Introduction

In the human body, there are several hormones that regulate different functions, such as growth and development, metabolism, electrolyte balance, and reproduction [1]. These functions have already been widely studied, and most of them are very well elucidated; even so, there are still aspects that have not yet been well explored. For example, the relationship between erythrocytes and hormones. Erythrocytes were treated as merely hemoglobin carriers for a long time, with small or none of the complex functions as seen currently.

It has already been seen that in healthy erythrocytes, adenosine can increase 2,3-bisphosphoglycerate (2,3-BPG) through activation of the ADORA2B receptor (A2B), suggesting that increasing adenosine levels increases oxygen release, which is positive for preventing tissue damage from acute ischemia. However, in sickly erythrocytes, the increase in oxygen release through the induction of 2,3-BPG by adenosine can be harmful, as oxygen release of oxygen induces the sickling of these erythrocytes [2].

In 1999, a study by Tuvia et al. showed that exposure of erythrocytes to adrenaline led to a concentration-dependent increase in erythrocyte deformability, and consequently, increased oxygen delivery to tissues [3]. Another study reported that adrenaline and epinephrine dose-dependently increase the rate of erythrocyte agglutination through alpha-1 adrenergic receptor activation. Furthermore, this study suggested that the effect of adrenaline was caused by an increase in Ca^{2+} entry into the erythrocyte, with consequent activation of the erythrocyte calmodulin, cyclooxygenase, and phospholipase A2 and leading to the release of K^+ from erythrocytes through Ca^{2+}-dependent K^+ channels, which is considered a manifestation of eryptosis. In summary, the potentiation of α1AR activation increases, while β2AR decreases the rate of eryptosis [4].

These studies present modulation of erythrocyte function by surface receptors and some of the signaling pathways that might be triggered inside the cell. In this chapter, the focus is to explore the actions of three well-studied hormones on erythrocyte function and some reports regarding intracellular pathways involved in these modulations.

2. Endothelin-1

Endothelins are peptide hormones, composed of 21 amino acid residues, capable of performing autocrine and paracrine functions. There are three distinct subtypes of endothelin, endothelin-1 (ET-1), endothelin-2 (ET-2), and endothelin-3 (ET-3). Endothelin-1 and endothelin-2 bind to endothelin receptor A (ETA) and endothelin receptor B (ETB), which are G-protein-coupled receptors, whereas endothelin-3 has a lower affinity for the ETA subtype. Endothelin-1 is considered a potent sub-nanomolar vasoconstrictor in the human cardiovascular system [5].

The vasoconstrictor mechanism of action was obtained from a culture medium of bovine aortic endothelial cells, described and characterized for the first time in 1985 by Hickey et al. It was from a potent contraction derived from an endogenous vasoactive peptide of the vascular endothelium [6]. A few years later, Yanagisawa and his group, in 1998, isolated endothelin to investigate its potential as a vasoconstrictor. Concluding with their research, endothelin, as well as neurotoxins, act directly on membrane ion channels, suggesting that the action of endothelin is closely associated with the influx of Ca^{2+}, through Ca^{2+} channels dependent on dihydropyridine, a calcium channel blocker that acts on smooth muscle cell [7].

Endothelins are synthesized into preprohormones and transformed posttranslationally into active peptides (**Figure 1**). Endothelin-1, which has been extensively studied, is synthesized with 212 amino acid residues (preproET-1), which are cleaved by an endopeptidase into Big-Et-1 (proET1), with 39 amino acid residues. This proET1 is in turn cleaved by endothelin converting enzyme (ECE), resulting in the active peptide hormone with 21 amino acid residues that play an important role in physiology [8].

Some works associate protein disulfide isomerase (PDI), based on evidence that shows the presence of PDI in the membrane of human erythrocytes, with the activity of the Gardos channel. Furthermore, a study by Prado and colleagues in 2013 showed that in the presence of endothelin-1, PDI activity increased, through a mechanism that includes casein kinase II. The Gardos channel is a K^+ channel activated by Ca^{2+}, in erythrocytes, being this channel related to homeostasis, so when this channel is activated, the cell dehydrates, leading to cell disorder and cell death. Prado suggests, as part of his research, the use of endothelin-1 receptor antagonists as a therapeutic target for sickle cell disease [9].

The ETB receptor has also been described in murine erythrocytes, and its presence has already been suggested in human erythrocytes. Foller and Rivera carried out studies showing the effect of endothelin-1 on erythrocytes regulation of Gardos channel activity, and this effect is due to the ET-1 binding to the ETB receptor. This effect interferes with the dehydration of sickled erythrocytes with a protective effect on the programmed death of murine erythrocytes (**Figure 2**) [10, 11].

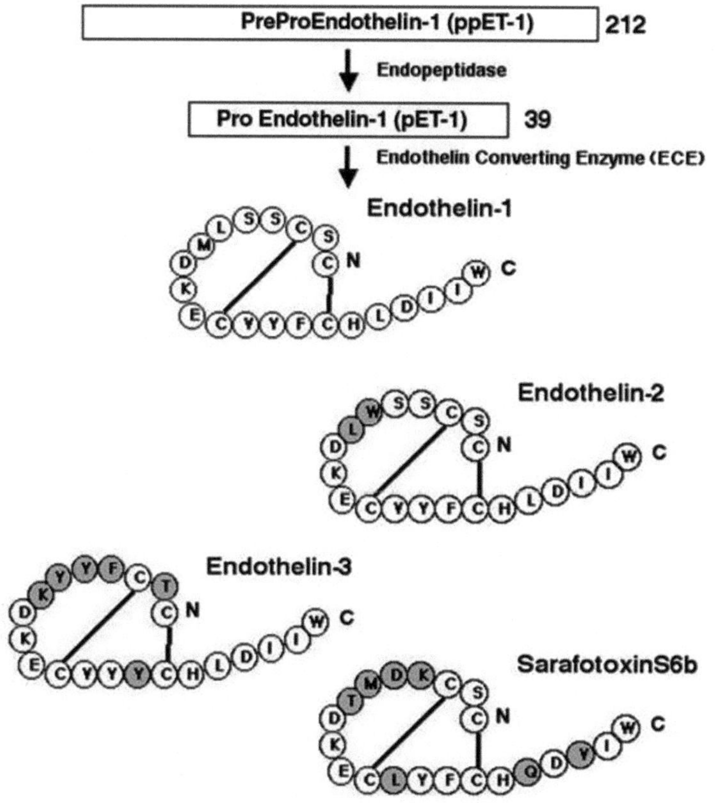

Figure 1.
Biosynthesis and amino acid sequence and structure of endothelin-1, endothelin-2, and endothelin-3 and related sarafotoxins. ET-2 and ET-3 differ from ET-1 by two and five amino acids, respectively, while sarafotoxin differs by seven amino acids [8].

A study carried out by George et al. in 2013 demonstrated the presence of the erythrocyte ETB receptor on the membrane of healthy and sickle cell anemia patients, associating the increased production of reactive oxygen species (ROS) to the presence of endothelin-1, concluding that these ROS were generated from erythrocyte NADPH oxidase (**Figure 3**). They demonstrated the oxidation of HbS, induced damage to the structure of red blood cells, increasing lysis, deformation, and vaso-occlusive process in a patient with sickle cell anemia. The inflammatory process that is increasing due to these processes alters plasma proteins, leukocytes, and endothelial cells, impairing the inflammatory condition. In addition, extracellular signaling molecules associated with this entire process end up acting back on erythrocytes *via* cell surface receptors, activating signaling pathways, such as the PKC and Rac pathways [12].

Prado in 2013 also showed a relationship between increased protein disulfide isomerase (PDI) activity and ET-1. Researchers performed experiments using BERK mice, which are mice mutated for sickle cell anemia, where they showed that reducing PDI activity improved hematological parameters and it was also possible to notice that there was modulation in the effects of ET-1 against the Gardos channel, further suggesting that this modulation occurred *via* the ETB receptor, which is present on the erythrocyte membrane [9]. These findings corroborate the study by Rivera in 2002, which showed that ET-1 induced changes in red blood cell volume and increased K^+

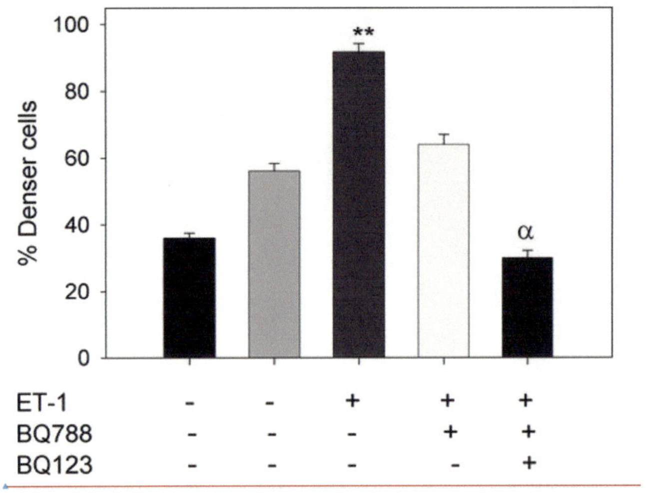

Figure 2.
Effects of ET-1 and its receptor antagonists on dehydration of oxygenated and non-oxygenated sickle cells [10].

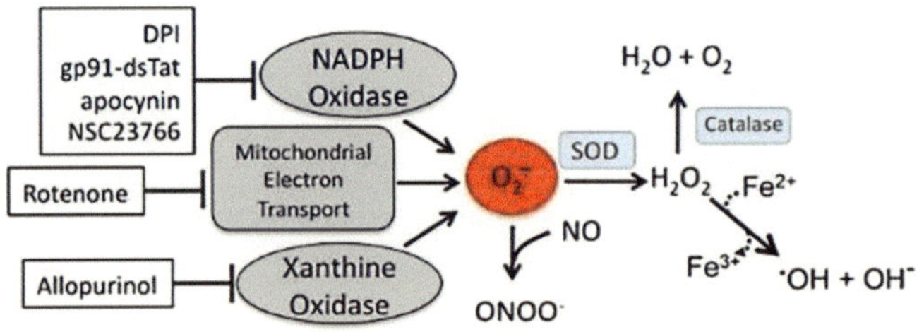

Figure 3.
Representation of nitric oxide synthase, ROS metabolism, and the presence of ROS enzyme inhibitors [12].

flux *in vivo.* Activation of endothelin receptors on healthy erythrocytes regulates the site of Ca^{2+} affinity or an undefined Ca^{2+}-dependent regulatory protein related to the Gardos channel. In the same study, they investigated whether this modulation would occur *via* the ETB receptor, using the receptor's antagonist, BQ-788, and showed a decrease in ET-1-induced activation of Gardos channel, both in healthy erythrocytes and in sickle erythrocytes [13].

Although the literature demonstrated the presence of ETB in erythrocytes, it is not well documented whether ETA is present and whether this receptor also modulates cell physiology. It is also important to compare the expression levels of these receptors in disease conditions compared to healthy counterparts to position the importance of endothelins in the disease process. Furthermore, it can be concluded that this endothelin-erythrocyte relationship is a powerful target therapeutic option in cases of anemia and also chronic pain, as reported by Smith et al. [14].

3. Thyroid-stimulating hormone (TSH)

Thyroid-stimulating hormone activity in the pituitary gland was first reported in the 1920s. However, it was not until the 1980s that the hormone thyrotropin (THS) had a detailed description of its structure [15, 16]. TSH is a glycoprotein produced by the thyrotrophs of the anterior pituitary gland. Thyrotropin-releasing hormone (TRH) is responsible for stimulating the synthesis and secretion of TSH, and inhibition is done through negative feedback by thyroid hormones (triiodothyronine and thyroxine—T3 and T4) [15, 17].

TSH, acting through the thyroid-stimulating hormone receptor (thyrotropin receptor—TSHR), is a G-protein-coupled receptor with seven transmembrane domains, and this protein can be Gq or Gs [18, 19]. Gs activities are mainly mediated by increased adenylate cyclase (AC) activity, which generates an increase in intracellular cAMP. This increase leads to direct activation of protein kinase A (PKA), activating CREB, or PKA activating the family pathway Ras. The Gq pathway mediates the activation of phospholipase C (PLC) and the Gβγ subunit, which will participate in the activation of second messengers, activating the Ras/Ras/Mek/Erk or PI3K/Akt pathways (**Figure 4**) [18–20].

In 2007, Balzan et al. identified TSHR in human erythrocyte membranes by Western blot (**Figure 5**). After the identification of TSHR, new research began on which pathways this hormone would act on the erythrocyte. In 2009, Balzan et al., demonstrated that TSH binds to TSHR in erythrocytes and modulates Na^+/K^+-ATPase, suggesting a new signaling pathway [21]. In 2020, Mendonça-Reis [22] showed that TSH at different concentrations (1–5 mIU/L) improved the resistance of red blood cells to hemolysis and this effect was caused by inhibition of the AMPK-dependent pathway and concomitant activation of the signaling pathway PI3K/Akt (**Figure 6**).

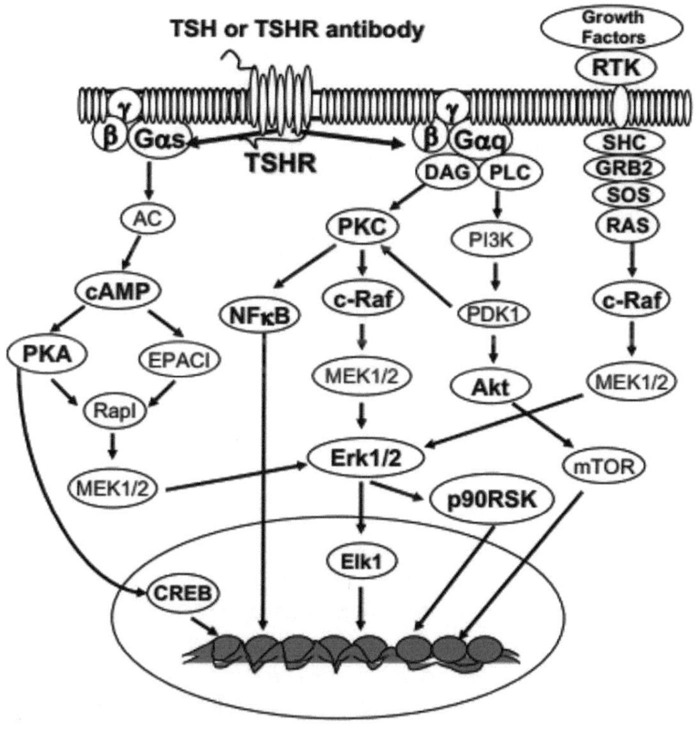

Figure 4.
Simplified illustrative diagrammatic image of the main signaling pathways involved with the G protein. Five pathways can be seen from left to right: cAMP/PKA/ERK or PKA/CREB, PI3/Akt/mTOR, PKC/NFB, PKC/c-raf/ERK/p90RSK, and Ras/c-Raf/ERK [18].

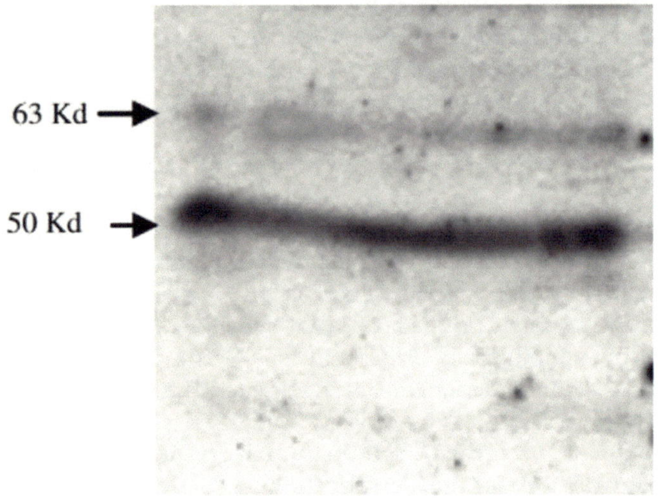

Figure 5.
Identification of TSH receptor on human erythrocyte membranes by Western blot [21].

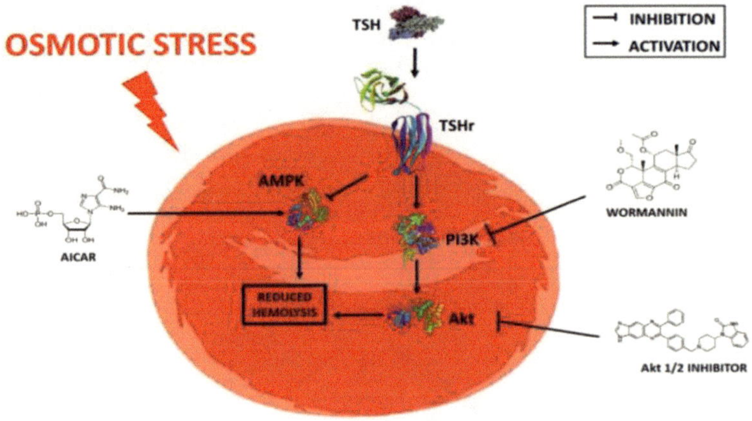

Figure 6.
Illustrative image demonstrating that TSH improves erythrocyte resistance to hemolysis in a situation of induced-osmotic stress, TSH binds to its receptor (TSHr) on the erythrocyte membrane, inhibiting the AMPK-dependent pathway (stimulated by AICAR) and concurrently activating the PI3K (inhibited by wortmannin)/Akt (inhibited by Akt 1/2 inhibitor) signaling pathway [23].

These new studies also opened a new range of research involving these receptors and diseases. For example, in the study by Ref. [24], it was suggested that erythrocyte Na^+/K^+-ATPase would be a good biochemical marker for subclinical hypothyroidism, as it is sensitive to subtle changes in thyroid function in this situation [24].

Patients with sickle cell disease (SCD) have also been found to have clinical hypothyroidism and high concentrations of TSH (6.4 mIU/L) [25]. And in ElAlfy et al. [26] observed impaired thyroid microcirculation and decreased thyroid volume among patients with SCD, and these factors were related to disease duration, but the results were not related to thyroid function, suggesting that these disorders can happen independently of the accumulation of iron [26].

On this basis, what can be observed from the identification of a functional TSH receptor in erythrocytes and some clarified pathways is that this hormone can modulate cell behavior and fate. All of these studies demonstrate in several ways the importance of continuing to elucidate the functions of TSH and its receptor on erythrocytes and how they may be involved in the pathophysiology of several diseases and serve as indicators of physiological changes.

4. Angiotensin II

The renin-angiotensin aldosterone system is an endocrine system responsible for regulating normal blood pressure *in vivo*, through the regulation of extracellular volume, vascular structure, and integrity, maintaining homeostasis [27, 28]. One of its main components is angiotensin II (Ang II), a vasoactive octapeptide responsible for several actions in tissues [29].

Two subtypes of Ang II receptors have been identified in humans, ATR1 and ATR2, described as receptors formed by seven transmembrane domains coupled to the G protein (**Figure 7**), with ATR1 being the dominant subtype and being widely distributed in the endocrine, renal, cardiac, and nervous systems [30]. While ATR2, after birth, has low expression, persisting only in some organs, such as the brain, kidney, and peripheral vasculature, mediating the physiological effects of Ang II, and may increase its expression in pathological conditions, such as hypertension, cardiac, and renal failure [28, 31].

There are studies about signaling pathways and the action of angiotensin in different cell types, however, there is little information in the literature about the actions of Ang II receptors on erythrocytes and their signaling, especially considering human erythrocytes.

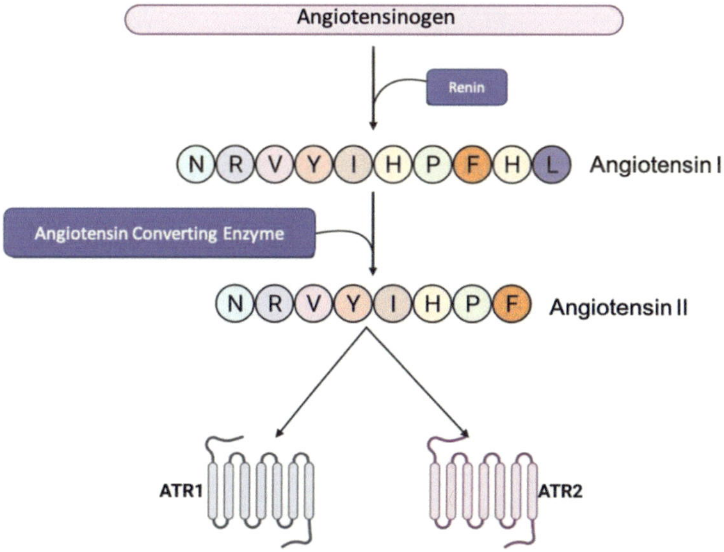

Figure 7.
Angiotensin II synthesis. Angiotensinogen is converted into angiotensin I by renin that in turn is converted into angiotensin II by the angiotensin converting enzyme. Angiotensin II can bind to ATR1 and ATR2.

In vitro experiments, carried out in 1997, showed a stimulatory effect of Ang II on erythropoiesis when erythroid progenitors were cultured in the presence of erythropoietin, demonstrating the importance of erythropoietin in erythropoiesis. Erythropoietin is an essential hormone for the regulation of erythropoiesis [32]. They also observed that the use of losartan (ATR1 antagonist, used to prevent vasoconstriction and volume expansion induced by circulating Ang II) blocked this stimulatory effect, demonstrating that Ang II, *via* the AT1 receptor, is responsible for mediating this stimulation [33, 34].

Reinforcing previous results in 2005, it was demonstrated that persistent activation of the renin-angiotensin system increases erythropoiesis in mice, *in vivo*, and that the most important receptor subtype responsible for this erythropoiesis was the AT1 receptor [35]. In 2015, in addition to reinforcing the idea that erythropoiesis and blood pressure are negatively regulated by inhibiting the ATR1 receptor, it was possible to say that the signaling pathways involved are complex and distinct since erythropoiesis is more resistant to inhibition of the ATR1 receptor than blood pressure control [30].

Some studies on the Ang II signaling pathway in erythrocytes have been described in order to elucidate the mechanism of the parasitic invasion of *Plasmodium falciparum* in erythrocytes. Research published in 2011 demonstrated that human erythrocytes express different Ang II receptors, namely ATR1, ATR2, and MAS receptors [36].

The levels of bradykinin (BK) and Ang- (1–7) increase in the presence of captopril (ECA inhibitor) in the supernatant of infected erythrocytes, decreasing parasite invasion and reducing PKA activity, through the association of receptors B2/MAS [37]. Thus, inhibition of protein kinase A (PKA) by the MAS receptor appears to be favorable to the erythrocyte against parasitic invasion.

Experiments in mice and humans showed that the activation of the ADORA2B-AMPK cascade, in the presence of angiotensin II, increased BPG mutase and consequently 2,3BPG, being beneficial for renal hypoxia, injury, proteinuria, and reduction of chronic kidney disease (CKD) [38].

In 2021, Guimares-Nobre and collaborators demonstrated that AT1 receptors and AT2 receptors, expressed in human erythrocytes, are capable of responding to osmotic stress situations in the presence of Ang II and its antagonists, losartan (ATR1 antagonist) and PD123319 (ATR2 antagonist). The study was carried out using modulators of signaling pathways, already recognized as activated by Ang II, in other cell types. As a result, it was possible to observe that, in osmotic stress, Ang II binds to the ATR2 receptor and reduces hemolysis through the PI3K/AKT and P38 pathways. However, when binding to ATR1, this protection did not occur (**Figure 8**) [39].

Furthermore, *ex vivo* experiments displayed that sickle cell erythrocytes treated with Ang II present increased cell deformability, decreased phosphatidyl serine translocation to the outer layer, and decreased hemoglobin S polymerization. All these factors are crucial findings for sickle cell disease outcomes (**Figure 9**) [40]. However, despite knowing this important action of Ang II in sickle cell erythrocytes, the precise mechanism of actions has not yet been demonstrated.

There is very little information regarding the importance of angiotensin II receptors on erythrocytes and the signaling pathways are triggered by it; however, researchers have been demonstrating that these receptors are functional and interfere in the survival of erythrocytes. Therefore, more studies are needed to expand this knowledge.

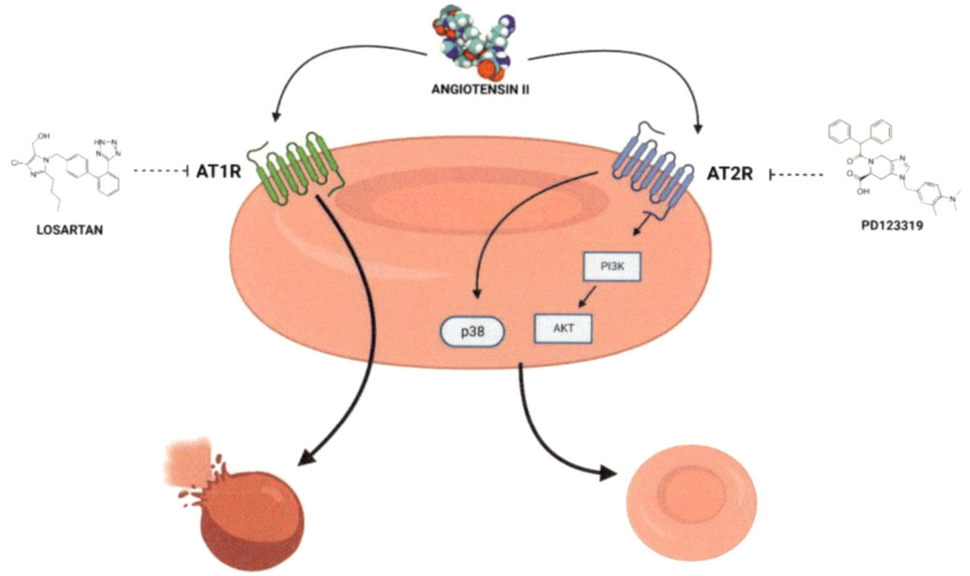

Figure 8.
Representative scheme of angiotensin II signaling pathway in erythrocytes in osmotic stress [39].

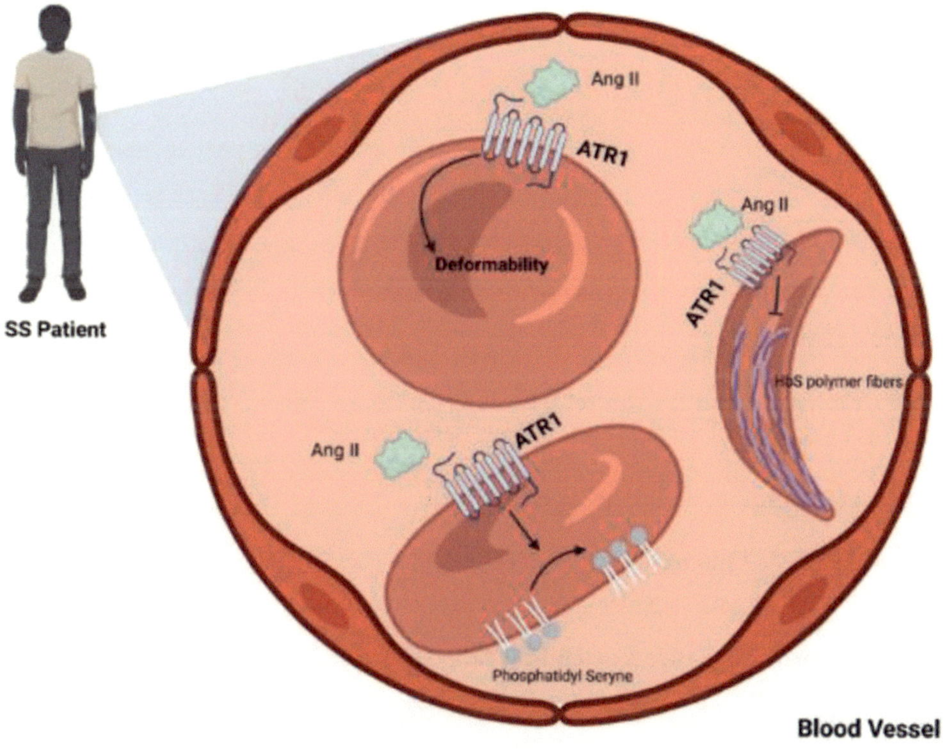

Figure 9.
Representative scheme of angiotensin II effects on sickle cell disease erythrocytes [39].

Author details

Camila Cristina Guimarães-Nobre[1,2], Evelyn Mendonça-Reis[1,2],
Lyzes Rosa Teixeira-Alves[1,2] and Clemilson Berto Junior[1,2]*

1 Universidade Federal do Rio de Janeiro, Brazil

2 Programa de Pós-Graduação em Endocrinologia, Faculdade de Medicina, UFRJ, Brazil

*Address all correspondence to: clemijr@yahoo.com.br

References

[1] Hiller-Sturmhöfel S, Bartke A. The endocrine system. Alcohol Health and Research World. 1998;**22**:153-164

[2] Zhang Y, Xia Y. Adenosine signaling in normal and sickle erythrocytes and beyond. Microbes and Infection. 2012;**14**:863-873

[3] Tuvia S, Moses A, Gulayev N, Levin S, Korenstein R. β-Adrenergic agonists regulate cell membrane fluctuations of human erythrocytes. The Journal of Physiology. 1999;**516**:781-792

[4] Volodchenko AI, Tsirkin VI, Kostyaev AA. The mechanism of change in the rate of agglutination of human erythrocytes under the influence of adrenaline. Human Physiology. 2014;**40**:171-178

[5] Davenport AP, Hyndman KA, Dhaun N, Southan C, Kohan DE, Pollock JS, et al. Endothelin. Pharmacological Reviews. 2016;**68**:357-418

[6] Hickey KA, Rubanyi GABOR, Paul RJ, & Highsmith RF. Characterization of a coronary vasoconstrictor produced by cultured endothelial cells. American Journal of Physiology-Cell Physiology. 1985;**248**(5):C550-C556

[7] Yanagisawa M et al. A novel potent vasoconstrictor peptide produced by vascular endothelial cells. Nature. 1988;**332**(6163):411-415

[8] Fagan KA, McMurtry IF, Rodman DM. Role of endothelin-1 in lung disease. Respiratory Research. 2001;**2**(2):1-12

[9] Prado GN. Endothelin-1 receptor antagonists regulate cell surface-associated protein disulfide isomerase in sickle cell disease—Prado – 2013. The FASEB Journal. 2013. Wiley Online Library

[10] Rivera A. Reduced sickle erythrocyte dehydration in vivo by endothelin-1 receptor antagonists. American Journal of Physiology-Cell Physiology. 2007;**293**:C960-C966

[11] Föller M, Mahmud H, Qadri SM, Gu S, Braun M, Bobbala D, et al. Endothelin B receptor stimulation inhibits suicidal erythrocyte death. The FASEB Journal. 2010;**24**:3351-3359

[12] George A, Pushkaran S, Konstantinidis DG, Koochaki S, Malik P, Mohandas N, et al. Erythrocyte NADPH oxidase activity modulated by Rac GTPases, PKC, and plasma cytokines contributes to oxidative stress in sickle cell disease. Blood. 2013;**121**:2099-2107

[13] Rivera A, Jarolim P, Brugnara C. Modulation of Gardos channel activity by cytokines in sickle erythrocytes. Blood. 2002;**99**:357-363

[14] Smith TP, Haymond T, Smith SN, Sweitzer SM. Evidence for the endothelin system as an emerging therapeutic target for the treatment of chronic pain. Journal of Pain Research. 2014;7:531-545

[15] Grossmann M, Weintraub BD, Szkudlinski MW. Novel insights into the molecular mechanisms of human thyrotropin action: Structural, physiological, and therapeutic implications for the glycoprotein hormone family. Endocrine Reviews. 1997;**18**:476-501

[16] Magner JA. Thyroid-stimulating hormone: Biosynthesis, cell biology, and bioactivity. Endocrine Reviews. 1990;**11**:354-385

[17] Green ED, Baenziger JU. Asparagine-linked oligosaccharides on lutropin, follitropin, and thyrotropin. I. Structural elucidation of the sulfated and sialylated oligosaccharides on bovine, ovine, and human pituitary glycoprotein hormones. The Journal of Biological Chemistry. 1988;**263**:25-35

[18] Michalek K, Morshed SA, Latif R, Davies TF. TSH receptor autoantibodies. Autoimmunity Reviews. 2009;**9**:113-116

[19] Morshed SA, Latif R, Davies TF. Characterization of thyrotropin receptor antibody-induced Signaling cascades. Endocrinology. 2009;**150**:519-529

[20] Mariotti S, Beck-Peccoz P. Physiology of the hypothalamic-pituitary-thyroid Axis MDText.com, Inc. 2021

[21] Balzan S, Del Carratore R, Nicolini G, Forini F, Lubrano V, Simili M, et al. TSH induces co-localization of TSH receptor and Na/K-ATPase in human erythrocytes. Cell Biochemistry and Function. 2009;**27**:259-263

[22] Evelyn M-R, Camila Cristina Guimarães N, Artur P-C, Leandro M-A, Clemilson B-J. Effects of TSH on erythrocyte osmotic fragility: Signaling pathway. International Journal of Blood Research Disorders. 2020;**2020**:7

[23] Mendonça-Reis E, Mendonça-Reis E. Effects of TSH on erythrocyte osmotic fragility: signaling pathway. International Journal of Blood Research and Disorders. 2020;7(1):1-7

[24] Nicolini G, Balzan S, Colzani R, Scarlattini M, Taddei MC, Iervasi G. Erythrocyte Na/K-ATPase is increased in subjects with subclinical hypothyroidism. Clinical Endocrinology. 2004;**60**:705-710

[25] Garadah TS, Jaradat AA, Alalawi ME, Hassan AB. Hormonal and echocardiographic abnormalities in adult patients with sickle-cell anemia in Bahrain. Journal of Blood Medicine. 2016;**7**:283-289

[26] ElAlfy MS, El-Sherif NH, Sakr HM, El Ashkar MNM. Thyroid hemodynamic alterations in Egyptian patients with sickle cell disease: Relation to disease severity, total body iron and thyroid function. Expert Review of Hematology. 2019;**12**:589-596

[27] Azushima K, Morisawa N, Tamura K, Nishiyama A. Recent research advances in renin-angiotensin-aldosterone system receptors. Current Hypertension Reports. 2020;**22**:22

[28] Karnik SS, Unal H, Kemp JR, Tirupula KC, Eguchi S, Vanderheyden PML, et al. International Union of Basic and Clinical Pharmacology. XCIX. Angiotensin receptors: Interpreters of pathophysiological Angiotensinergic stimuli [corrected]. Pharmacological Reviews. 2015;**67**:754-819

[29] dos Santos AF, Almeida CB, Brugnerotto AF, Roversi FM, Pallis FR, Franco-Penteado CF, et al. Reduced plasma angiotensin II levels are reversed by hydroxyurea treatment in mice with sickle cell disease. Life Sciences. 2014;**117**:7-12

[30] Kato H, Ishida J, Matsusaka T, Ishimaru T, Tanimoto K, Sugiyama F, et al. Erythropoiesis and blood pressure are regulated via AT1 receptor by distinctive pathways. PLoS One. 2015;**10**

[31] Savoia C, Burger D, Nishigaki N, Montezano A, Touyz RM. Angiotensin II and the vascular phenotype in hypertension. Expert Reviews in Molecular Medicine. 2011;**13**:e11

[32] Tsiftsoglou AS. Erythropoietin (EPO) as a key regulator of

erythropoiesis, bone remodeling and endothelial transdifferentiation of multipotent mesenchymal stem cells (MSCs): Implications in regenerative medicine. Cell. 2021;**10**:2140

[33] Michel MC, Brunner HR, Foster C, Huo Y. Angiotensin II type 1 receptor antagonists in animal models of vascular, cardiac, metabolic and renal disease. Pharmacology & Therapeutics. 2016;**164**:1-81

[34] Mrug M, Stopka T, Julian BA, Prchal JF, Prchal JT. Angiotensin II stimulates proliferation of normal early erythroid progenitors. The Journal of Clinical Investigation. 1997;**100**:2310-2314

[35] Kato H, Ishida J, Imagawa S, Saito T, Suzuki N, Matsuoka T, et al. Enhanced erythropoiesis mediated by activation of the renin-angiotensin system via angiotensin II type 1a receptor. The FASEB Journal. 2005;**19**:2023-2025

[36] Saraiva VB, de Silva LS, Ferreira Silva CT, da Silva Filho JL, Teixeira-Ferreira A, Perales J, et al. Impairment of the plasmodium falciparum erythrocytic cycle induced by angiotensin peptides. PLoS One. 2011;**2011**:6

[37] de Silva LS, de Peruchetti DB, Silva CT, Ferreira-Silva AT, Perales J, Caruso-Neves C, et al. Interaction between bradykinin B2 and Ang-(1-7) Mas receptors regulates erythrocyte invasion by Plasmodium falciparum. Biochimica et Biophysica Acta. 2016;**1860**:2438-2444

[38] Peng Z, Luo R, Xie T, Zhang W, Liu H, Wang W, et al. Erythrocyte adenosine A2B receptor-mediated AMPK activation: A missing component counteracting CKD by promoting oxygen delivery. JASN. 2019;**30**:1413-1424

[39] Guimarães-Nobre CC, Mendonça-Reis E, Passinho-da-Costa L, Miranda-Alves L, Berto-Junior C. Signaling pathway in the osmotic resistance induced by angiotensin II AT2 receptor activation in human erythrocytes. rbmb.net. 2021;**10**:314-326

[40] Guimarães-Nobre CC, Mendonça-Reis E, Teixeira-Alves LR, Miranda-Alves L, Berto-Junior C. ATR1 angiotensin II receptor reduces Hemoglobin S polymerization, phosphatidylserine exposure, and increases deformability of sickle cell disease erythrocytes. Cell Biochemistry and Biophysics. 2022;**80**:711-721

Chapter 4

Reticulocytes-Mother of Erythrocytes

Ashish Kumar Gupta and Shashi Bhushan Kumar

Abstract

Reticulocytes are immature red blood cells (RBCs) that is seen in the bone marrow after through nuclear extrusion from the orthochromatic normoblasts. They are released into the peripheral blood as mature RBCs, after completion of maturation in the bone marrow. The reticulocyte count reflects the erythropoietic activity of the bone marrow, the rate of reticulocyte delivery from the bone marrow into the peripheral blood, and the rate of reticulocyte maturation. Reticulocyte enumeration is also of value in monitoring bone marrow regenerative activity after chemotherapy or bone marrow transplantation. Manual counting of reticulocytes by light microscopy with supravital dyes for RNA remains the standard method of reticulocyte enumeration. However, automated methods of reticulocyte enumeration developed during the past decade are much more accurate, precise, and cost-effective than manual counting, and are increasingly being performed in the clinical laboratory. The differentiation of the reticulocyte is based on the presence of RNA. The newer techniques provide a variety of reticulocyte related parameters, such as the reticulocyte maturation index and immature reticulocyte fraction, which are not available with light microscopy. These new parameters are under evaluation in the clinical diagnosis and monitoring of hematological disorders.

Keywords: reticulocyte count, new methylene blue, immature reticulocyte fraction, reticulocyte maturation index, reticulocyte production index, high-fluorescence ratio, medium-fluorescence ratio, low-fluorescence ratio, reticulocyte mean hemoglobin content

1. Introduction

Reticulocytes are immature red blood cells (RBCs) produced in the bone marrow and released into the peripheral blood where the terminal maturation into RBCs occurs for next couple of days. Any alteration in reticulocyte count is an indicator of active or failed erythropoiesis, in response to anemias or other causes of bone marrow dysfunction [1].

The first description of reticulocytes was made in 1865 by Wilhelm Heinrich Erb, a German neurologist when he discovered the population of granulated erythrocytes while observing the effect of acetic and picric acid on the development of erythrocytes. He erroneously regarded these cells as transitional forms between leucocytes and erythrocytes. Ehrlich described the stained material as fine, dense, and elegant networks as a feature of senescent erythrocytes rather than of young ones [2]. The reticular substance was first regarded as a degenerative material, a "coagulation

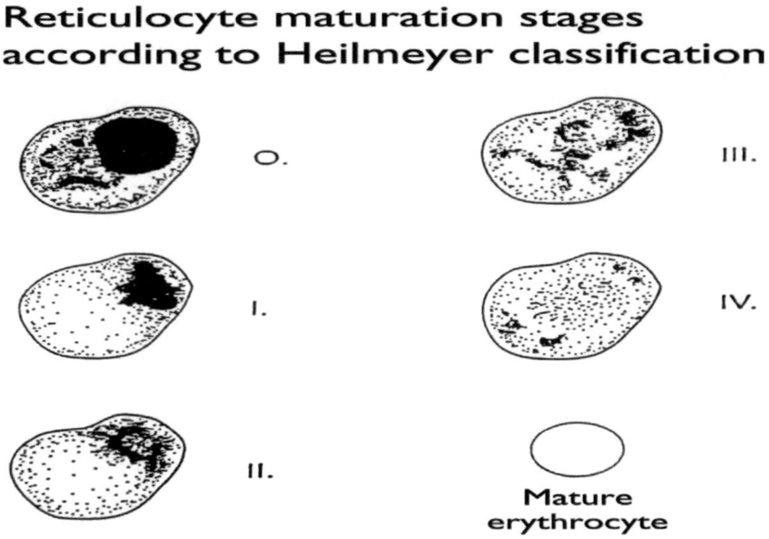

Figure 1.
Maturation stages of reticulocytes according to Heilmeyer classification [4].

necrosis" or a substance produced by the action of certain deleterious agents on corpuscles.

Theobald Smith brilliantly asserted that reticulocytes represented young red cells. Further investigators designated these cells as erythrocytes with "substantia granulo-filamentosa," the Americans used such terms as "reticulated red cell" or "vital-staining erythrocyte."

In 1922, Edward Bell Krumbhaar, first coined the term "reticulocyte" when he stated that: "Erythrocytes revealing granular filamentous substance by the methods of vital staining may be conveniently designated 'reticulocytes" [3].

Ludwig Heilmeyer, a German internist, proposed the still well-known classification of reticulocytes maturity in 1932 (**Figure 1**). In 1944, Pierre Dustin, a Belgian pathologist, showed that granular "reticulated" substance is RNA [5]. In 1947, Giovanni Astaldi, an Italian hematologist, classified reticulocytes into 3 stages of maturity, as in the current maturation classification used in flow cytometry. In 1918, it was demonstrated that reticulocytes are able to synthesize hemoglobin and absorb iron [6].

Over the years acridine orange, a fluorescent dye, replaced the use supravital stain, improving the sensitivity of manual microscopic count of reticulocytes [7].

2. Definition of reticulocytes

2.1 Origin of the term and definition

The "reticulocyte" term has derived from the reticulum of RNA and protein precipitated by the fixation and staining that seen microscopically after supravital staining [8, 9]. These supravital granules represent precipitated rough endoplasmic reticulum along with polyribosomes. Thus, the human reticulocytes are defined as mature red cells that form a reticulum network or granules on exposure to supravital stains, such as new methylene blue or brilliant cresyl blue [10].

Although this definition has been revised several times, in 1997 it was chosen to be the universal standard by the Clinical and Laboratory Standards Institute and the International Council for Standardization in Hematology (NCCLS-ICSH) [11].

2.2 Morphology

Romanowsky type stain, such as May-Gru¨nwald Giemsa or Wright stain, causes the RNA to disappear during alcohol fixation and reticulocytes acquire slightly larger size than mature erythrocytes, with a uniform polychromatic blue-gray color. On staining with new methylene blue (NMB) or brilliant cresyl blue (BCB), the RNA precipitates and becomes visible with the characteristic scattered granules under the microscope.

Reticulocytes must have at least 2 blue staining granules, visible without fine microscope adjustment and present away from the cell margin, as standard morphological definition provided by NCCLS-ICSH, in 1997 [11]. These granules can mimic Heinz bodies.

2.3 Properties of reticulocytes

Reticulocytes in comparison with mature erythrocytes, have following characteristics [12]:

- The early reticulocyte contains mitochondria, a small number of ribosomes, the centriole, and remnants of Golgi bodies.

- Early reticulocytes continue to synthesize hemoglobin, with approximately 20–30% of the total hemoglobin of the RBC.

- Reticulocytes are larger (8.5 μ) than mature erythrocytes and they gradually decrease in size during maturation.

- The volume of reticulocytes and mature erythrocytes increases in pathological conditions but the ratio of their volumes remain constant: approximately 1.24

- Reticulocytes are less dense than mature RBCs, due to lower Hb concentration and higher water content of reticulocytes.

3. Reticulocyte maturation

When the orthochromatic normoblast (late stage of erythropoiesis) loses its nucleus, it becomes a reticulocyte that persist in the bone marrow for next few days and is subsequently released into the circulation for terminal maturation [13, 14].

Maturation of reticulocyte is a continuous process with various morphologic, biochemical, and functional changes that lead to remodeling of membrane, changes in volume, and elimination of membrane-bound organelles and ribosomes [15]. Immature or early reticulocytes are biochemically more active than mature ones with intact cellular functions of Hb production and absorption of iron [16]. Circulating reticulocytes are unable to synthesize Hb and cannot further increase their Hb content.

Maturation of reticulocytes is a complex sequential mechanism of enucleation, caused by condensation of chromatin, vesicular trafficking, and selective autophagy [17]. The ultimate maturation occurs when the basophilic reticular filamentous substance in the reticulocyte disappears [18]. Intracytoplasmic organelles such as the mitochondria, ribosomes, and endosomal vesicles are eliminated by a mitochondrial death program which includes physiologic events of macroautophagy and mitoptosis [19, 20].

Early reticulocyte maturation is characterized by the selective elimination of unwanted plasma membrane proteins (CD71, CD98, and β1 integrin) through the endosome exosome pathway. In contrast, late maturation is characterized by the generation of large glycophorin A coated vesicles of autophagic origin [21, 22].

Recent studies have suggested that the small amount of RNA that remains in reticulocytes might still be essential for reticulocyte maturation to form normal biconcave erythrocytes [23].

During the maturational remodeling of the membrane cytoskeleton, by vesiculation and endocytosis, reticulocytes lose about 24% of their volume and surface area, and increase their stability and deformability [24].

As already mentioned, a series of progressive physiological and biochemical changes occur during the differentiation of reticulocytes into mature RBCs [25]. The most important of these changes include:

1. Synthesis of hemoglobin and its cytoplasmic accumulation.

2. Loss of protein-synthesizing apparatus and mitochondria.

3. Condensation and contraction of chromatin to cause extrusion of the nucleus.

4. Exosome formation to cause loss of cell-surface membrane receptor expression.

5. Alteration in cholesterol and phospholipid levels of cell membrane.

6. Changes in intracytoplasmic enzyme levels of glucose-6-phosphate dehydrogenase (G-6-PD).

4. Classification of reticulocytes

A number of classification systems have been attempted based on the maturation of reticulocytes along with their morphology. Maturity classification is based on the quantity of the granular/reticular filamentous substance and its distribution in their cytoplasm.

The first attempt was made by Heilmeyer and Westharer, who divided the cells into 4 groups (Groups I-IV), designated by Roman numerals and characterized by a progressive reduction in the compactness of the reticulum. Therefore, reticulocyte maturation predominantly assesses the relative proportion of mRNA (see **Figure 1**).

Group 0 – nucleated RBCs with a dense perinuclear reticulum.

Group I – most immature reticulocytes with extruded nuclei having a dense, coherent mass of RNA.

Group II – reticulocytes with extensive but loose reticular network.

Group III – reticulocytes with scattered reticular network.

Group IV – most mature reticulocytes with scattered remnants of RNA.

According to Lowenstein, in steady-state erythropoiesis the circulating reticulocytes are more than 60% in Group IV, 30% in Group III, and less than 1% in Groups I or II.

4.1 Stress Reticulocytosis

Some specific hematological disturbances (*e.g.* severe anemia or hemolysis) are compensated with accelerated erythropoiesis (called stress erythropoiesis), which causes release of the immature, larger and more stained form of reticulocytes into the peripheral blood, which are called "stress reticulocytes" and appear as polychromatophilic cells on Romanowsky staining. All polychromatophils are reticulocytes on a Wright-stained blood smear. Reticulocyte production can increase up to 20 folds above the base line values of 1 to 2 million reticulocytes per second when stressed with intense hematopoiesis. This increased production is accomplished by increased production and shortening of marrow maturation time [26, 27]. Stress reticulocytosis is seen in bone marrow regeneration following autoimmune hemolysis, chemotherapy induced anemia, administration of therapy in nutritional anemia, and use of erythropoiesis-stimulating agents (ESAs). Stress reticulocytes are multilobular and motile, in contrast to the cup-shaped non-motile mature reticulocytes.

5. Reticulocyte counting

Till the end of last century, the standard method of counting reticulocytes was based on the visual detection of RNA ribosomal networks (reticulum), by microscopic examination of supravitally stained peripheral blood smears. But, this method has several sources of imprecision in the manual microscopic counting of reticulocytes, including different staining with variation, distributional variability for quality of blood film, intra and inter-observer variations, and inadequate number of cells counted. As per NCCLS-ICSH guidelines, supravital staining with New Methylene Blue (NMB) still remains the recommended method for optimal agreement/correlation assessments. NMB staining procedures have been standardized to improve the accuracy and reliability of reticulocyte manual counting [28].

Reticulocyte counting should be done within 6 hours if the sample is kept at room temperature, or up to 72 hours if the blood sample is refrigerated at 2-6°C.

5.1 Manual methods

5.1.1 Photographic methods

In 1960s photographs used to be obtained on wet preparation from fresh oxalated blood mixed with brilliant cresyl blue in isotonic saline. 30–35 fields containing reticulocytes used to be printed on glossy paper with final magnification of approximately 4000x.

5.1.2 Planimetric method

Reticulocytes stained with brilliant cresyl blue wet preparations are photographed and the resultant Kodachromes are projected onto butcher paper. Then, pairs of reticulocytes and adjacent RBCs are circled and traced with a planimeter.

5.1.3 Light microscopic method

Equal volumes of anticoagulated (preferably EDTA) peripheral whole blood and supravital stains are mixed and incubated for at least 20 minutes. A thin smear of the stained blood preparation is made on a microscope slide, a counterstain (usually Wright's) is applied, and the slide is examined by light microscopy. An adequate number of RBCs (minimum 1000 for optimal analytical precision) in a well-stained area are examined, and the proportion of reticulocytes is determined. Reticulocytes possess a blue granular precipitate, which can vary from individual small blue granules to a network of blue reticular material. Reticulocytes have a faint, diffuse basophilic hue termed as polychromasia (**Figure 2**).

The reticulocyte count is usually reported as a percentage of total RBC count. The normal mean percentage reticulocyte count by NMB light microscopy is 1.0% to 3.0%. In cases of anemia, this relative reticulocyte count is misleading because reduced RBC count causes erroneous elevation in reticulocyte count. Under these circumstances, reticulocyte count is corrected with respect to patient's packed cell volume (PCV) as Reticulocyte index to compensate the decrease in mature RBCs. The PCV corrects the percentage reticulocyte count to the baseline PCV i.e. 0.45 using the following formula:

$$\text{Reticulocyte Index}(RI) = \text{Reticulocyte count}(\%) \times \frac{\text{Patient's}\,PCV}{0.45}$$

Absolute count of reticulocytes is more accurate term to correct the effect of anemia. This is calculated as follows:

$$\text{Absolute Reticulocyte count}(ARC) = \text{Reticulocyte Count}(\%) \times RBC\ \text{count}\left(per\ mm^3\right)$$

The normal ARC is between 50,000 and 150,000 reticulocytes/mL (5×10^{10} and 1.5×10^{12} reticulocytes/L).

Manual microscopic RC is highly subjective and tedious, which results in high level of imprecision, with a coefficient of variations of up to 50% [30]. This imprecision in manual counting of reticulocytes can be attributed to:

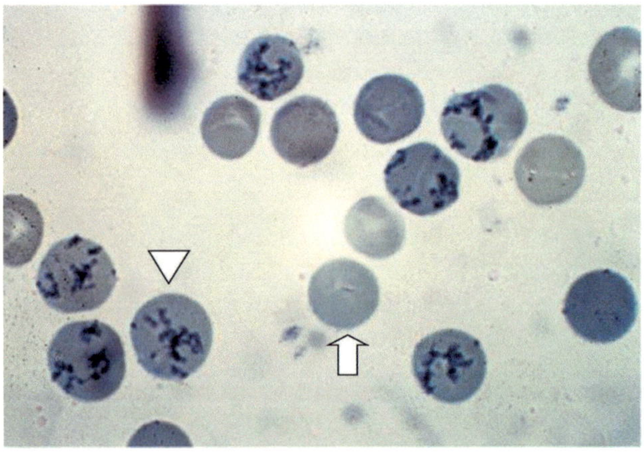

Figure 2.
Photomicrograph of peripheral blood smears stained with new methylene blue, 400x. Reticulocytes (arrowhead) are differentiated from mature RBCs (arrow) by the presence of an intracellular granular precipitate [29].

- Different supravital stains (NMB or BCB)
- Staining variations
- Quality of blood film
- Intra and interobserver variation
- Number of reticulocytes and RBCs counted
- Stain precipitates and cellular interferences (platelets, leucocytes fragments)
- Erythocytic inclusions (Heinz bodies, Basophilic stippling etc.)

The College of American Pathologists and National Committee for Clinical Laboratory Standards (NCCLS) has defined reticulocytes as the cell containing two or more precipitate granules not attached to cytoplasmic membrane following staining by new methylene blue and the use of Miller ocular disks, to standardize the microscopic counting and reduce the imprecision.

Hence, clinical serial assessment of erythropoietic activity of bone marrow in patients receiving myelotoxic or hematinic therapies by this manual counting of reticulocytes is impractical due to poor precision and methodologic limitations.

5.2 Immunofluorescence method

First attempt of reticulocyte enumeration by using immunofluorescence microscope was done in 1950s by Kozenow and Mai. They used RNA specific fluorochrome dye i.e., acridine orange. Subsequently, different fluorochromes used by others for reticulocyte enumeration later on like pyronin Y, thioflavin T, DiOCl, proflavin. Reticulocyte enumeration by this method gives spuriously high count sometimes due to some interference present in the blood sample such as platelet clumps, nucleated RBCs, Howell-Jolly bodies, Heinz bodies, Pappenheimer bodies, Basophilic stippling and Malarial parasites.

5.3 Fluorescence microscopy

The prefixed whole blood in formalin is mixed with a dilute, buffered solution of the dye and counted as a wet preparation. Ultraviolet to blue light is required for the excitation of these dyes, and a green long-pass filter must be placed for detection of the fluorescence emission.

5.4 Conventional flow cytometry

Thioflavin T analogue has been widely used as specific fluorochrome for reticulocyte enumeration to improve counting accuracy and made flow cytometric reticulocyte enumeration practical as TO excitation occurs in the visible region of the spectrum. These flow cytometers rely on the enumeration of reticulocytes using nucleic acid binding fluorescent dyes and measurement of fluorescence emission. Reticulocytes are identified and enumerated within the gated RBC population on the basis of fluorescence intensity. The quantitation of the intensity of green fluorescence reveals young, immature reticulocytes as brightly fluorescent (high RNA content), while maturing reticulocytes show an intermediate fluorescence intensity (intermediate RNA content), and older reticulocytes show dim fluorescence (low RNA content).

5.5 Automated reticulocyte instrumentation

All automated instruments for reticulocyte enumeration are capable of rapid analysis of whole blood sample in flow through systems, with the red cell population interrogated on a cell-by-cell basis by laser light. Now-a-days the reticulocyte counting is fully integrated into the automated complete blood count (CBC) in high-throughput hematology analyzers. The reticulocyte RNA content is detected by fluorochrome Auramine O, while the cell size is determined by forward light scattering using an argon ion laser as the light source. Automated flow cytometric analysis has replaced traditional microscopic quantitation of reticulocytes.

5.6 Reticulocyte-specific monoclonal antibodies

Reticulocyte analysis with fluorochrome-labeled monoclonal antibodies against reticulocyte surface receptors is available only for research studies of the reticulocyte and diseases of the erythron. Fluorochrome-labeled monoclonal antibodies specific for different reticulocyte antigens have been evaluated as reagents for reticulocyte enumeration. The antigens presently evaluated include the CD71 molecule (transferrin receptor, TfR, and T9 antigen), CD36, and other antigens. CD71 receptors are abundant on the surface of early reticulocytes, but gradually decrease during reticulocyte maturation, and are absent from mature RBCs. CD36 and the integrin $\alpha4$-$\beta1$ complex are expressed on reticulocytes but not mature erythrocytes. CD36-positive reticulocytes correlate with the stress reticulocyte fraction in patients with chronic hemolytic diseases.

6. Indices of reticulocytes

The recent development in automation of reticulocyte counts provide accurate RC with enhanced precision along with reliable and accurate measurements of RNA content and cellular indices such as mean reticulocyte volume, mean reticulocyte Hb concentration, and its content. These novel parameters have been studied with prompt interest for their clinical usefulness and the utility of reporting these analytes with their appropriate interpretation.

7. Reticulocyte production index (RPI)

During intense erythropoietic stimulation, immature, large and basophilic precursor macroreticulocytes ("shift cells") from the bone marrow are released prematurely into the peripheral blood. This causes a shortened reticulocyte maturation time in the bone marrow (depends on the severity of stress erythropoieisis), and the longer reticulocyte maturation time in the peripheral blood. Since these shift cells have a cell volume about 25% larger than that of normal cells, a correction for RBC maturation time and the PCV must be done when they comprise more than 5% of the total reticulocytes [31]. The correction is referred as "shift correction" or "reticulocyte production index", and calculated by the following formula (**Figure 3**).

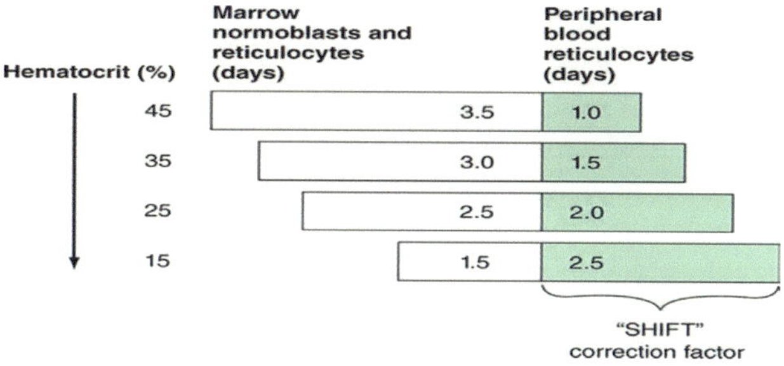

Figure 3.
The relationship between PCV (hematocrit) and blood maturation time. The total RBC maturation time is approximately 4.5 days. During normal erythropoieisis 3.5 days of maturation occurs in bone marrow and last one day in peripheral blood. However, in anemia the marrow maturation time progressively shortens, and reticulocytes circulate for a correspondingly longer period of time in the peripheral blood to compensate the degree of anemia [32].

$$\text{Reticulocyte Production index}\,(RPI) = \frac{\text{observed reticulocyte count}\,(\%) \times \dfrac{\text{Patient's}\ PCV}{0.45}}{\text{maturation time in peripheral blood}\,(\text{Shift Factor})}$$

RPI was proposed to correct the reticulocyte percentage for peripheral blood maturation time based on the hematocrit value. It helps to alleviate the effect of the premature release of reticulocytes by taking into account maturation time of reticulocytes. Thus RC is corrected or adjusted for both premature release of reticulocytes and the degree of anemia. Stressed erythropoiesis is accomplished by increased production and shortening of the fraction of time that reticulocytes mature in marrow and proportionally prolongs their maturation time in circulation, thus increasing reticulocyte circulation time. RPI is used for evaluating erythropoiesis and classifying anemias. RPI is normally between 2 and 3.

RPI of less than 2 suggests for hypo-proliferative erythropoiesis and more than 3 is applied for hypo-proliferative state of erythropoiesis.

Clinical interpretation of RPI

- Increased RPI (RPI >3)
  - Hemolytic anemias (Autoimmune)
  - Recent hemorrhage
  - Marrow response to therapy in nutritional anemia (EPO)
  - Chemotherapy induced anemia
- Decreased RPI (RPI < 2)
  - Hypoproliferative disorder (i.e., aplastic anemia)
  - Ineffective erythropoiesis
  - Megaloblastic anemia

8. Reticulocyte maturity index (RMI)

Different populations of reticulocytes depend on the content of RNA in the enucleated cytoplasm. This RNA content can be assessed by the different intensity signals of fluorescence or light scattering/absorbance obtained on the flow cytometric based automated hematoanalyzers. The software-based algorithm in the autoanalyzers discriminate these reticulocytes into three areas of clusters according to stain intensity. High-fluorescence light scatter/absorbance reticulocytes correspond to young or immature reticulocytes, whereas maturing reticulocytes have medium-fluorescence light scatter, and older reticulocytes have low fluorescence light scatter. These three classes are called high-fluorescence ratio (HFR), medium high-fluorescence ratio (MFR), and low-fluorescence ratio (LFR), respectively.

The RMI is directly proportional to the amount of reticulocyte intracellular RNA. Maturational stages of reticulocytes in peripheral blood depends upon the level of anemia, disease state and iron status. Thalassemia, megaloblastic anemia, anemia of uremia and myelodysplastic syndrome (MDS) are associated with delayed reticulocyte maturation. Hence, RMI is used as adjunct parameter in the evaluation of hematological disorders and therapeutic monitoring of erythropoietic activity. Responsive marrow is expected to manifest a high RMI along with a subnormal RC.

9. Immature reticulocyte fraction (IRF)

Immature reticulocyte fraction is better accepted term internationally to quantify the younger fraction of reticulocyte as a sum of HFR and MFR, to avoid the ambiguity in the interpretation of different fluorescence intensities of RMI. It is expressed as a fraction (0.00–1.00). IRF assesses reticulocyte maturation by the intensity of the staining that reflects the mRNA content. Assessment of IRF is clinically useful when it is evaluated in correlation with absolute RC.

Clinical utility of IRF:

- **Early marker of engraftment in bone marrow**- IRF increases very early when the bone marrow engraftment is successful and it precedes other parameters, like absolute neutrophil count (ANC), reticulated platelets, and RC. Increase in IRF occurs up to 1 week sooner than increase in ANC. IRF value greater than 10% indicates early marrow recovery and more than 20% from the post BMT value suggests successful erythroid engraftment. Hence, serial determination of IRF after bone marrow transplantation (BMT) is used to demonstrate successful engraftment.

- **Early marker for stem cell mobilization**- IRF has been proposed as an early marker of $CD34^+$ mobilization for peripheral stem cell harvesting to optimize the timing for stem cell collection following growth factor mobilization or cytotoxic drug therapy.

- **Effective monitoring of therapeutic efficacy of erythropoietin therapy**- IRF increases in response to treatment with ESAs before there is an increase of RC in renal failure, AIDS, and MDS. Thus, indicate adequate erythropoietic stimulation.

- **Monitoring for efficacy of anemia treatment**- As the IRF increases earlier than the reticulocyte number, it is useful in monitoring the efficacy of therapy in nutritional

Disease/State	ARC	RMI	IRF
Aplastic anemia	L	L	L
Aplstic crisis	L	L/N	L
Hypoplastic anemia	L	N/H	L
Bone marrow regeneration	L	N/H	H/N
Chronic disease	L/N	N	N
Iron deficiency	L/N	H	H
Thalassemia	N/H	N/H	N/H
MDS	Any Level	N/H	N/H
Megaloblastic anemia	L/N	H	H
Hemolytic anemia	H	H	H
Hemorrhage	N/H	H	H

L: Low, N: Normal, H: High, ARC: Absolute Reticulocyte Count, RMI: Reticulocyte Maturity Index, IRF: Immature Reticulocyte Fraction.

Table 1.
Diagnostic utility of reticulocyte indices [34].

anemias. In ineffective erythropoiesis, IRF is increased while reticulocyte count is reduced or normal, in some cases of MDS or in dyserythropoietic anemia.

- **Screening of Hereditary Spherocytosis:** High RC without proportionate elevation in IRF can be suspicious for HS. A RC/IRF ratio higher than 7.7 can be used as a cut off for the screening of all HS cases as a diagnostic algorithm [33].

The clinical utility of IRF has also been reported in a variety of conditions, like: the monitoring of anemia treatment and neonatal transfusion needs; monitor response to EPO in a blood conservation program; renal transplant engraftment from Epo production; the detection of occult or compensated hemorrhages or hemolysis, and aplastic crisis in hemolytic anemias; and the early diagnosis and monitoring of aplastic anemias (**Table 1**).

10. Reticulocyte mean cell Hemoglobin content (CHr)

As understood the early reticulocytes continue to synthesize hemoglobin, with approximately one fourth of the total hemoglobin of the RBC. Hence, CHr is the measurement of the Hb content of reticulocytes expressed in pg./cell. CHr is the product of the cellular volume and the cellular Hb concentration. The measurement of CHr can directly reflect the functional availability of iron in that time frame of life span of reticulocyte (up to 4 days) and is helpful in real-time assessment of the functional state of erythropoiesis. It can be of help in identifying iron deficiency before the development of IDA and also its serum level can guide for the route of administration of iron supplementation. Critically low value of CHr warrants for intravenous iron therapy, because oral iron is ineffective in preventing iron-deficient erythropoiesis. The prediction of the absence of bone marrow iron stores can better be diagnosed by CHr less than 28 pg., over than MCV, serum ferritin or transferrin saturation. Adequate iron levels must be ensured to optimize Hb production in a balance with erythropoietic stimulation by Epo in cases of chronic renal disease and it can be assessed by CHr.

11. Summary

The final stage of RBC differentiation occurs in the peripheral blood. The reticulocytes that are released by the bone marrow still contain RNA. On the way of maturation reticulocytes gradually lose their rough endoplasmic reticulum and mitochondria to become mature RBC after about 3–4 days in the peripheral blood. Since the number and characteristics of the reticulocytes in the peripheral blood gives insight about the activity of the bone marrow, reticulocyte counting has become fundamental part of the hematopoietic evaluation now-a-days. Circulating reticulocytes decrease in patients with impaired bone marrow function, and increase in cases of hemolysis with normal bone marrow activity. Reticulocyte enumeration by light microscopy, with the use of a supravital dye (*viz.* NMB), which binds to the RNA in the reticulocyte still remains the standard method for RC. However, the accuracy and precision of this assay are greatly compromised by its subjective nature. In contrast, automated techniques of reticulocyte enumeration are more precise, accurate, objective, and cost-effective. A variety of RNA-specific fluorescent dyes have been utilized for automated reticulocyte enumeration, and some hematology analyzers utilize optical light scatter analysis to perform reticulocyte analysis on specimens stained with NMB or other dyes. In addition to relative and absolute reticulocyte counts, automated techniques provide information regarding the age distribution of reticulocytes in form of the RMI, IRF, CHr. There is extensive evidence that these parameters are useful in the accurate classification of anemia patients, and monitoring patients receiving EPO or recovering from chemotherapy or bone marrow transplantation. The recent trend to incorporate reticulocyte analysis into the routine hematology analyzer has made automated reticulocyte analysis increasingly such common that, perhaps in recent future, the manual reticulocyte count will become an obsolete technique.

Author details

Ashish Kumar Gupta[1*] and Shashi Bhushan Kumar[2]

1 AIIMS, Raipur, India

2 Sub Divisional Hospital, Pusa, Samastipur, India

*Address all correspondence to: pathimsbhu@yahoo.com

References

[1] Riley RS, Ben-Ezra JM, Goel R, Tidwell A. Reticulocytes and reticulocyte enumeration. Journal of Clinical Laboratory Analysis. 2001;**15**(5): 267-294

[2] Ehrlich P. Ubereinige Beobachtungen am ana¨mischen Blut. Berl Klein Wschr. 1881;**18**:43

[3] Krumbhaar EB, Chanutin A. Studies on experimental plethora in dogs and rabbits. Experimental Biology and Medicine. 1922;**19**:188-190

[4] Piva E, Brugnara C, Spolaore F, Plebani M. Clinical utility of reticulocyte parameters. Clinics in Laboratory Medicine. 2015;**35**:133-163

[5] Dustin P. Contribution a` l'e'tude histophysiologique et histochimique des globules rouges des verte'bre's. Archives de Biologie. 1944;**55**:285-292

[6] London IM, Shemin D, Rittenberg D. Synthesis of heme in vitro by the immature non-nucleated mammalian erythrocytes. The Journal of Biological Chemistry. 1950;**183**:749-755

[7] Vander JB, Harris CA, Ellis SR. Reticulocyte counts by means of fluorescence microscopy. The Journal of Laboratory and Clinical Medicine. 1963;**62**:132-140

[8] Seip M. Reticulocyte studies. Acta Medica Scandinavica. 1953;**282**(Suppl):9-164

[9] Houwen B. Reticulocyte maturation. Blood Cells. 1992;**18**:167-186

[10] Brecher G. New methylene blue as a reticulocyte stain. American Journal of Clinical Pathology. 1949;**19**:895

[11] Koepke JA, Broden PN, Corash L, et al. NCCLS document H44-A-Methods for Reticulocyte Counting (Flow Cytometry, and Supravital Dyes); Approved Guideline. 1997;**17**(15)

[12] d'Onofrio G, Chirillo R, Zini G, et al. Simultaneous measurement of reticulocyte and red blood cell indices in healthy subjects and patients with microcytic and macrocytic anemia. Blood. 1995;**85**:818-823

[13] Skadberg O, Brun A, Sandberg S. Human reticulocytes isolated from peripheral blood: Maturation time and hemoglobin synthesis. Laboratory Hematology. 2003;**9**:198-206

[14] Koury MJ, Koury ST, Kopsombut P, et al. In vitro maturation of nascent reticulocytes to erythrocytes. Blood. 2005;**105**:2168-2174

[15] Bertles J, Beck W. Biochemical aspects of reticulocyte maturation. The Journal of Biological Chemistry. 1962;**237**:3770-3777

[16] London IM, Shemin D, Rittenberg D. Heme synthesis and red blood cell dynamics in normal humans and in subjects with polycythemia vera, sickle-cell anemia, and pernicious anemia. The Journal of Biological Chemistry. 1949;**179**:463-484

[17] Sandoval H, Thiagarajan P, Dasgupta SK, et al. Essential role for nix in autophagic maturation of erythroid cells. Nature. 2008;**454**:232-235

[18] Chen M, Sandoval H, Wang J. Selective mitochondrial autophagy during erythroid maturation. Autophagy. 2008;**4**:926-928

[19] Shimizu S, Honda S, Arakawa S, et al. Alternative macroautophagy and mitophagy. The International Journal of Biochemistry & Cell Biology. 2014;**50**:64-66

[20] Geminard C, de Gassart A, Vidal M. Reticulocyte maturation: Mitoptosis and exosome release. Bio Cell. 2002;**26**:205-215

[21] Ney PA. Normal and disordered reticulocyte maturation. Current Opinion in Hematology. 2011;**18**:152-157

[22] Griffiths RE, Kupzig S, Cogan N, et al. Maturing reticulocytes internalize plasma membrane in glycophorin A-containing vesicles that fuse with autophagosomes before exocytosis. Blood. 2012;**119**:6296-6306

[23] Lee E, Choi HS, Hwang JH, et al. The RNA in reticulocytes is not just debris: It is necessary for the final stages of erythrocyte formation. Blood Cells, Molecules & Diseases. 2014;**53**:1-10

[24] Liu J, Guo X, Mohandas N, et al. Membrane remodeling during reticulocyte maturation. Blood. 2010;**115**:2021-2027

[25] Palis J, Segel GB. Developmental biology of erythropoiesis. Blood Reviews. 1998;**12**:106-114

[26] Manwani D, Bieker JJ. The erythroblastic island. Current Topics in Developmental Biology. 2008;**82**:23-53

[27] Chasis JA, Mohandas N. Erythroblastic islands: Niches for erythropoiesis. Blood. 2008;**112**:470-478

[28] Arkin CF, Davis BH, Bessman BH, et al. NCCLS. Methods for Reticulocyte Counting (Automated Blood Cell Counters, Flow Cytometry, and Supravital Dyes); Approved Guideline. 2nd ed. NCCLS Document H44-A 2. Wayne (PA): NCCLS; 2004

[29] Dayyal Dg. How to count reticulocytes (manual method). Bioscience. 2018. ISSN 2521-5760

[30] Savage R, Skoog D, Rabinovitch A. Analytical inaccuracy and imprecision in reticulocyte counting: A preliminary report from the College of American Pathologists's reticulocyte project. Blood Cells. 1985;**11**:97-112

[31] Koepke JF, Koepke JA. Reticulocytes. Clinical and Laboratory Haematology. 1986;**8**:169-179

[32] Loscalzo J, Fauci AS, Kasper DL, Hauser SL, Longo DL, Jameson JL. Harrison's Principles of Internal Medicine. 21st Edition. McGraw Hill/ Medical; May 2022

[33] Mullier F, Lainey E, Fenneteau O, et al. Additional erythrocytic and reticulocytic parameters helpful for diagnosis of hereditary spherocytosis: Results of a multicentre study. Annals of Hematology. 2011;**90**:759-768

[34] Davis BH, Bigelow NC. Automated reticulocyte analysis: Clinical practice and associated new parameters. Hematology/Oncology Clinics of North America. 1994;**8**(4):617-630

Section 2

Erythrocytes - Diseases

Chapter 5

Genetics of Thalassemia

Poonam Tripathi

Abstract

β-Thalassemia is a common genetic disorder caused by mutations in β-globin gene that results in reduced β-globin production. There are more than 200 different mutations that have been reported till date affecting the diverse levels of β-globin gene expression and causing β-thalassemia. Nucleotide substitutions and frameshift insertion-/deletion-type mutations interfere with the molecular mechanism like transcription of the β-globin gene, splicing process and translation of mRNA of β-globin gene, thus resulting in either absence or reduction of synthesis of β-globin chains. Molecular analysis is a must for all thalassemia patients. Definitive diagnosis and counseling of these patients will help in better management of disease.

Keywords: genetics, thalassemia, hemoglobin, gene, mutation, severity

1. Introduction

1.1 Thalassemia

The thalassemia syndromes are inherited blood disorder caused by mutations in either alpha or beta gene that decreases the synthesis of either alpha or beta globin chains [1]. This results in ineffective erythropoiesis, hemolysis, and ultimately leading to a variable degree of anemia [2]. The main types of thalassemia reported on the basis of the type of globin chains are affected, grouped as α, β, δβ, γδβ, δ, γ and εγδβ thalassemia [3].

Developmental stages	Hemoglobin
Embryonic	Hb Grower1 ($\zeta_2\varepsilon_2$)
	Hb Grower2 ($\alpha_2\varepsilon_2$)
	Hb Portland 1 ($\zeta_2\beta_2$)
	Hb Portland 2 ($\zeta_2\gamma_2$)
Fetal	HbF (α2γ2)
Adult	HbA (α2β2
	HbA_2 (α2δ2)

Table 1.
Different types of hemoglobin at various developmental stages of human.

In the alpha (α)-thalassemia, there is reduced production or absence of α-globin subunits, whereas in the beta (β)-thalassemia, there is reduced production of β-globin subunits. The β-thalassemia can be clinically classified according to the degree of severity i.e., the beta-thalassemia carrier state, thalassemia intermedia and thalassemia major (TM) [4]. Thalassemia trait or thalassemia carrier type is asymptomatic one, whereas thalassemia intermedia is more severe form than trait and they occasionally need blood transfusions, while a person with thalassemia major is transfusion dependent severe form of disease and to sustain life need regular blood transfusions (**Table 1**) which later in life develop secondary complications like iron overload, subsequent tissue damage and oxidative stress [5, 6].

2. Structure of hemoglobin (Hb)

Hemoglobin is a tetramer composed of 2-alpha (α)-globin and 2-beta (β)-globin chains working in combination with heme to transport oxygen in the blood (**Figure 1**). It is iron containing protein, synthesized inside immature erythrocyte in the red bone marrow. The globin polypeptides bind heme molecule, which in turn allows the hemoglobin in erythrocytes to bind oxygen reversibly and transport it from the lungs to other part of body [7, 8].

The main hemoglobin type is HbA in normal adult constitute about 98% and contains two alpha and two beta globin chains (α2β2). The minor type HbA_2 constitute less than 3% of the adult hemoglobin consists of two alpha chains and two delta chains (α2δ2). HbF is predominant hemoglobin sub-type, found only during fetal life and consist of two alpha and two gamma (γ) subunits (α2γ2) [9, 10].

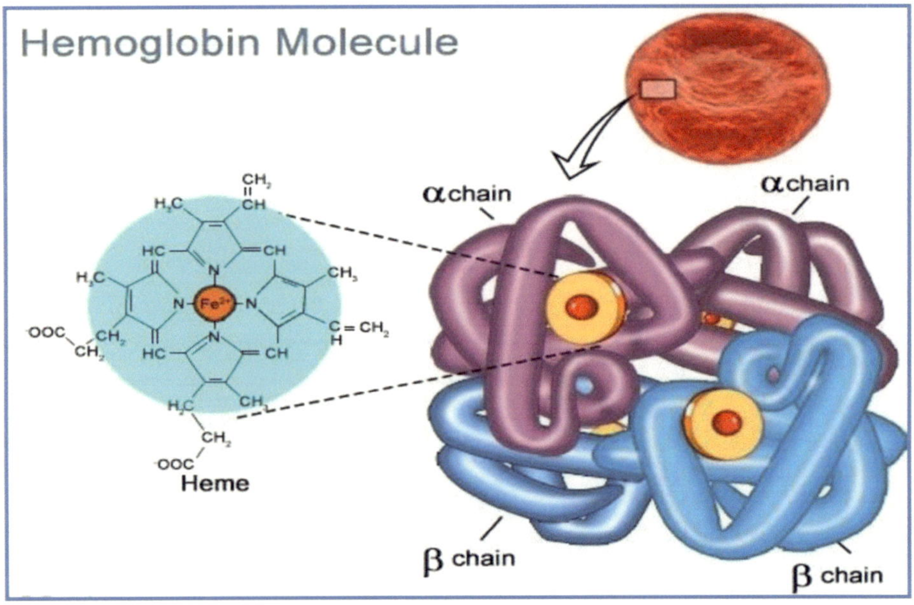

Figure 1.
Structure of hemoglobin [2].

3. Hemoglobin expression during embryogenesis

In vertebrates, there are typically several forms of Hb differing in composition of polypeptide chains and for oxygen transport under special conditions in various stages of human development. The developmental changes in Hb production are brought by differential activation of globin genes, which is largely determined at level of transcription. This process is known as hemoglobin switching [11].

During early stages of human development, embryonic Hb is expressed in RBC progenitors, which develops in the yolk sac. This Hb molecule consists of two hetero dimmers of epsilon (ε) and zeta (ζ)-globin chains [12]. The first switch in globin composition occurs as the site of erythropoiesis changes from the yolk sac to the fetal liver at 12 weeks post-conception period. Production of the embryonic form (Hb Portland-2 ($\zeta_2\beta_2$), Hb Portland-1 ($\zeta_2\gamma_2$), Hb Gower-1 ($\zeta_2\varepsilon_2$), Hb Gower-2 ($\alpha_2\varepsilon_2$) is downregulated, and it is replaced by fetal hemoglobin (HbF) containing α- and γ-globin chains (**Table 1**) [6, 13, 14] and by 17 weeks, only 1% of cells continue to express ζ-globin [15]. The second switch occurs approximately 6 weeks after birth: where the levels of fetal γ-globin decline and are replaced with adult hemoglobin (HbA). HbF makes up less than 1% of total hemoglobin in most individuals by 1 year of age [3]. The embryonic and adult α gene share 58% homology at amino acid level [16].

4. Genetic control of hemoglobin synthesis

The globin gene clusters show variability in their base composition and are organized into alpha globin gene cluster and beta globin cluster.

4.1 Alpha (α) globin gene cluster

The α-globin gene cluster is located on the short arm of Chromosome 16 (16p13.3). The cluster includes three protein coding functional genes (α_1, α_2, and ζ_2) and spans about 30 kb, and the genes are arranged in order as per their expression during developmental stages in human beings [17, 18].

The α1 and α2 gene are highly homologous and encode an identical protein and differs only in the sequences of non-coding regions [17, 19]. The α-globin gene codes for 141 amino acids and consist of 3 exons separated by two introns (**Figure 2**). The length of IVS 1 in both α 1 and α 2 gene is same having 117 nucleotides whereas IVS II of α1 consist of 149 and α2 is of 142 nucleotides.

4.2 Beta (β) globin gene cluster

The β-globin gene cluster is located along a 60-kb segment of the short arm of chromosome 11 (11p15.5) and contains five functional genes (β, δ, $^G\gamma$, $^A\gamma$, and ε) and one pseudogene (**Figure 3**). The genes are arranged in the order in which they are expressed during different developmental stages [21]. The regulatory sequences contain four (HS-1 to HS-5) erythroid specific DNAse hypersensitive sites (HSs) known as locus control region (LCR), which is required for DNA-protein interaction. The *HBB* gene, spans about 1.6 Kb, contains three exons (coding regions) and two introns with 5'and 3'untranslated regions (UTRs). The HBB gene, spans about

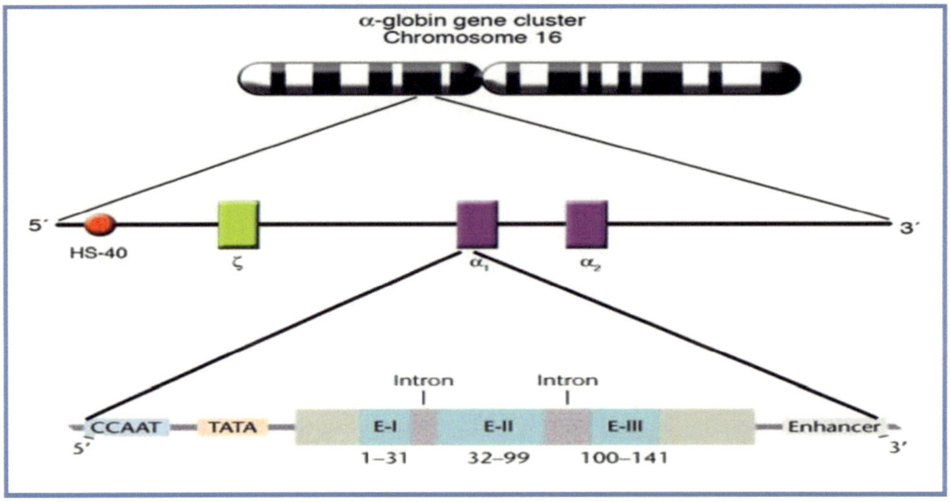

Figure 2.
Alpha globin gene cluster [20].

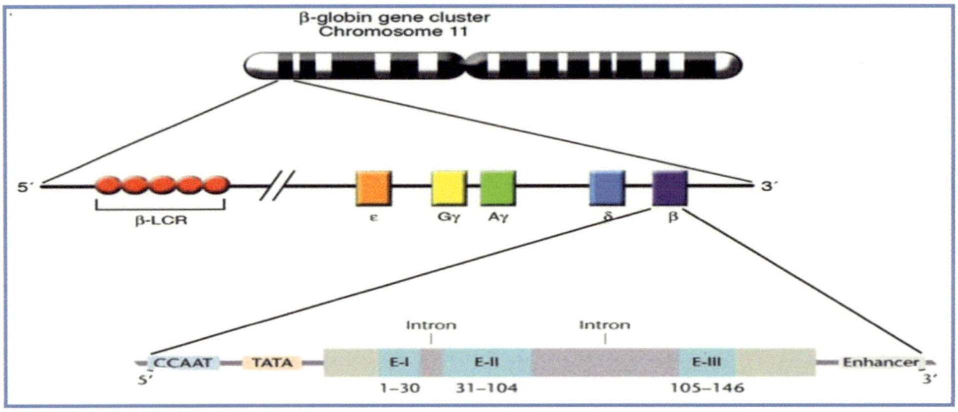

Figure 3.
Beta globin gene cluster [20].

1.6 Kb, consist of three exons, interrupted by two introns. The HBB gene is regulated by an adjacent 5′ promoter in which a TATA, CAAT, and duplicated CACCC boxes are located. A major regulatory region, containing a strong enhancer, 50 Kb from the beta globin gene [22]. This region, locus control region (LCR), contains four (HS-1 to HS-4) erythroid specific DNAse hypersensitive sites (HSs), which are involve in DNA-protein interaction. Each HS site is constituted by a combination of several DNA motifs interacting with transcription factors, among which the most important are GATA-1 (GATA indicates the relative recognition motif), nuclear factor erythroid 2, erythroid Krüppel-like factor [23]. The importance of LCR for the control of the beta-like globin gene expression has been discovered by studying a series of naturally occurring deletions that totally or partly remove the HS sites and result in the inactivation of the intact downstream beta globin gene. Several transcription factors bind and regulate the function of the HBB gene, the most important of which is erythroid Krüppel-like factor 1 (KLF 1), which binds the proximal CACCC box [21, 22].

Each gene cluster consist of the structural genes that are separated on both their 3′ and 5′ ends by variable stretches of non-coding DNA containing several types of regulatory sequences [23–25].

- Promoter elements are essential for the binding of messenger RNA polymerase and the initiation of transcription. HBB is immediately preceded by its promoter region. There are five main motifs in HBB's promoter region:
    - BRE at −37 GGGCTGG
    - TATA box at −31 CATAAAAG
    - Inr at +1 TTACATT
    - MTE at +27 ACAACTG
    - DPE at +32 AGCAA
- Enhancer and silencer elements stimulate or repress transcriptional activity depending on the array of proteinaceous "transcription factors" to which they are bound and which have two main motifs: the CAT BOX of the pattern CCAAT, and the CACC box of the pattern CACCC [26].
- The Locus Control Region (LCR) functions as a "master switch." Both globin gene clusters (alpha and beta globin) possess a LCR, located many kilobases upstream of the HBB and HBA gene. The LCR appears to interact with a combination of transcription factors at the onset of erythroid maturation in such a way as to enhance access of the transcriptional machinery and other transcriptional factors to the promoters, enhancers, and silencers within the gene complex. LCR function is absolutely required for expression of globin genes at the extraordinary high levels needed for normal hemoglobin synthesis [25].

5. Beta thalassemia mutation types

Thalassemia is characterized by decrease or absence production of α- or β-globin chains, which result in α- or β-thalassemia [25]. More than 400 different types of deletional and non deletional types of mutations in α- or β-globin genes have been reported till date with diverse clinical manifestations, ranging from asymptomatic to fatal anemia [26].

- β-Thalassemia (complete absence of hemoglobin subunit beta) alleles result from nonsense, frameshift, or splicing mutations.

- β^+-Thalassemia alleles are produced by pathogenic variants in the promoter area (either the CACCC or the TATA box), the polyadenylation signal, or the 5′ or 3′ untranslated region, or by splicing abnormalities.
- The complex β-thalassemia (delta-beta- and gamma-delta-beta-thalassemia) results from deletion of variable part of *HBB* gene cluster.
- β-Thalassemia also caused by deletion of the LCR regions of *HBBgene*.
- In rare cases of the β-thalassemia mutations reported beyond the β-globin gene cluster.

5.1 Gene deletions

- Deletion of one or more alpha globin genes is the most common mechanism responsible for alpha thalassemia in Asian and Mediterranean populations. On the other hand, complete deletion of the beta globin gene is rarely reported, as part of larger gene rearrangement lesions. Rare cases of beta thalassemia are caused by partial beta globin gene deletions and interstitial deletions.
- Large deletions within the gene clusters—The "classical" clinically harmless forms of Hereditary Persistence of Fetal Hb (HPFH) appear to arise from deletional events that remove large regions of DNA from the beta gene cluster. In these cases, 50–100 kilobase long segments of DNA are deleted from downstream regions of the beta globin gene cluster. These deletions remove the delta and beta globin structural genes [27].
- Persistence of very high levels of fetal hemoglobin synthesis in adult life appears to occur in those cases in which the gene deletion brings a highly active enhancer element into close proximity with the remaining gamma globin gene. This enhancer, usually insulated from the globin gene cluster, provides for high levels of gamma globin gene expression and persistence of Hb F in adult life [27, 28].
- Other deletions involving this region cause mild-to-moderate forms of thalassemia in which both delta and beta globin synthesis are absent (delta-beta thalassemia). The clinical pattern seems to depend on the size and location of the downstream deletions; each brings new DNA, with varying enhancer effects, in close proximity to the remaining genes of gamma [29].
- Deletions of LCR—these cases are very rare, but extremely informative forms of both beta and alpha thalassemia have been discovered to arise from gene deletion events that remove the locus control region (LCR) sequences. The classical mutation, reported in families, was associated with total absence of beta globin synthesis, even though the beta globin gene and its surrounding promoters and enhancers were found to be normal in structure [30]. The chromosome responsible for this form of thalassemia was found to have undergone a large deletion many thousand bases upstream of the structural genes, through which the critical LCR sequences were lost. Thus the LCR played critical role by permitting expression of the globin genes during erythropoiesis [31].

- These events are associated with biochemical and clinical features of hereditary persistence of fetal Hb (HPFH) with the added presence of the structural variant, Hb Kenya. The persistence of high levels of fetal hemoglobin synthesis into adult life may arise because of the removal of stage selector elements and silencers normally positioned between the gamma and delta genes, and the closer apposition of a strong beta globin gene enhancer normally located at the 3′ side of the beta gene to the G (gamma) and Kenya genes. Hb Kenya is rare; the anti-Kenya state has yet not been identified [32, 33].

5.2 Mutations affecting transcription

Alpha and beta thalassemia have been found to arise from mutations that alter known promoter or enhancer sequences for alpha or beta globin genes [28]. These point mutations alter the efficiency of the promoter or enhancer, while others are small gene deletions or rearrangement that disrupts their spatial integrity [29].

5.3 Mutations affecting Pre-mRNA splicing

Many mutations have been described that disrupt normal splicing of the mRNA precursor. Among these are some of the most common forms of beta thalassemia and more common varieties of "non-deletion" forms of alpha thalassemia [21, 34].

5.4 Alteration of canonical splice signals

Some thalassemic splicing mutations directly disrupt the canonical "splicing signals" used to mark the beginning and end of each intron so that normal splicing can occur. These short sequences are required by the splicing machinery. They signify the places at which excision of the intron, and ligation together of the flanking exons, should occur [34, 35].

Certain bases in these splicing signals are "invariant", such as the GT dinucleotide required at the 5′ beginning of the intron and the AG dinucleotide required at the 3′ end of the intron. The several bases to either side of these invariant nucleotides are consensus sequences within which alteration of the base will change the efficiency with which the site is used [36]. Thus, mutations altering these nucleotides can abolish normal splicing or reduce it to a variable degree. Mutations that alter the consensus splice sites reduce production of alpha or beta globin mRNA; the pre-mRNA molecules which are not properly spliced appear to be catabolized rapidly, so that abnormal mRNA products do not accumulate [37].

In some cases, pre-mRNAs that are not spliced at the proper sites are spliced elsewhere by activation of "cryptic" sites, resulting in the production of structurally abnormal, usually non-functioning messenger RNAs [35]. The molecular basis for variability with which mutations reduce the efficiency of normal splicing, or generate production of detectable abnormally spliced products, remains poorly understood and needs to be investigated.

5.5 Activation of cryptic splicing sites

Another class of splicing mutations, one of which produces a very common form of thalassemia in the Mediterranean region, includes those in which the mutation activates a "cryptic" splice site. These mutations are not located within the consensus sites at either end of introns. Rather, base substitution, small deletions, or small insertions of DNA, can convert a site within an exon or intron that normally bears only a slight resemblance to a splice site into one containing much stronger consensus signal [37, 38]. Depending upon the consequential "strength" of the signals, the splicing apparatus will utilize that site instead of normal site in a greater or lesser percentage of the pre-mRNA molecules being spliced. At least two spliced mRNA products result, accumulating in varying percentages, depending on efficiency with which the cryptic site is used and the stability of the abnormally spliced product [29, 38].

5.6 Altered mRNA translation and stability

Mutations that create a premature translation termination codon (nonsense codon) account for the most common forms of thalassemia, in terms of numbers of patients affected. These mutations create translation stop signals prematurely, so that the complete beta globin polypeptide is never made. In the most common type of thalassemia, globin fragments are highly unstable, resulting in the accumulation of protein synthesized from the mutated gene (i.e., beta (0) thalassemia) [38]. As a result, patients homozygous for this defect cannot make any beta chains and suffer from a severe form of beta thalassemia [27].

Premature translation termination can occur by base substitution or deletion of bases in an exon, producing so called "frame shift" mutations. The inserted or deleted stretch of DNA contains a number of bases that is an exact multiple of three, so "open" translation reading frame is maintained. So, the insertion or deletion will cause the ribosome to begin reading the codons out of the normal reading frame. One consequence of this "frame shift" mechanism is that the probability that a UAA, UAG, or UGA translation termination codon will be encountered with the shifted reading frame within 50 or so bases downstream [38, 39].

In the normal reading frame, these three bases (UAA, UAG, UGA) are usually divided among two codons, and thus never "read" as a stop codon by the ribosome. In the shifted reading frame, the three bases will appear as a single codon and translation ceases. Thus frame shifting results into premature translation termination. A reason for occurrences of both alpha and beta thalassemia mutations.

6. Nonsense-mediated decay

The physiologic consequences of translation termination are very clear; functional globin not synthesized and thalassemia results. Less obvious is the phenomenon that these prematurely terminated mRNAs accumulate in greatly reduced amounts. The reason behind impaired accumulation is not clear, since their transcription, processing, and stability elements are intact. Recent research work has shown that there are normal cell defense mechanisms to eliminate abnormally translated mRNAs. They probably exist in order to guard against the accumulation of truncated protein products, which have the potential to interact abnormally with other proteins and damage cells. This protective phenomenon is called nonsense-mediated decay.

Curiously, premature stop codons that occur in the final exon of either the alpha or beta globin gene accumulate at nearly normal levels. Moreover, the truncated polypeptides also accumulate stably and in significant amounts. It appears that the process of nonsense-mediated decay affects only those mRNAs in which the premature stop codon occurs in the first or second exons. This has been found in other gene systems as well. Indeed, there is increase in substantiation that nonsense-mediated decay and mRNA splicing are interactive processes [32].

7. Post translational mechanisms

A few, comparatively uncommon, mutations that cause thalassemia by disrupting the structure of the fully translated globin product. That appears to interfere with normal folding of the globin peptide to form stable dimers or tetramers. In each case, the abnormal globin generates inclusion bodies and produces a thalassemia phenotype. The final common pathway for these mutations is similar to that of the dominant form of thalassemia due to nonsense codons in the final exon. In all circumstances, accumulation and generation of substantial amounts of the abnormal protein results in formation of inclusion bodies [34, 35, 38].

8. Conclusion

Thalassemia arises from well over 400 mutations that, in the cumulative, affect every step required for successful production of the large amounts of Hb needed for normal red cell homeostasis. The mutations accounting for most of the thalassemia patients around the world are those affecting translation termination and mRNA splicing. Gene deletions and rearrangements, and defects affecting transcription, mRNA stability, or Hb assembly are uncommon or occur rarely.

Author details

Poonam Tripathi
Sanjay Gandhi Postgraduate Institute of Medical Sciences, Lucknow, India

*Address all correspondence to: poonam.tripathi90@gmail.com

References

[1] Schrier SL. Pathophysiology of thalassemia. Current Opinion in Hematology. 2002;**9**:123-126

[2] Vichinsky EP. Changing patterns of thalassemia worldwide. Annals of the New York Academy of Sciences. 2005;**1054**(1):18. Bibcode:2005NYASA1054...18V. DOI: 10.1196/annals.1345.003

[3] Thein SL, Menzel S, Lathrop M, et al. Control of fetal hemoglobin: New insights emerging from genomics and clinical implications. Human Molecular Genetics. 2009;**18**:R216-R223

[4] Weatherall DJ, Clegg JB. The Thalassaemia Syndromes. 4th ed. Oxford: Blackwell Science; 2001

[5] Weatherall DJ. The Thalassemias: Disorders of Globin Synthesis. Williams Hematology, 9e Eds. Kaushansky K, et al. McGraw Hill; 2015. Available from: https://accessmedicine.mhmedical.com/content.aspx?bookid=1581&sectionid=94305138

[6] Cao A, Galanello R. Beta-thalassemia. Genetics in Medicine. 2010;**12**(2):61-76

[7] Forget BG, Hardison RC. The normal structure and regulation of human globin gene clusters. In: Disorders of Hemoglobin: Genetics, Pathophysiology, and Clinical Management. 2nd ed. Cambridge University Press; 2009. pp. 46-61. DOI: 10.1017/CBO9780511596582.007

[8] Marengo-Rowe AJ. Structure-function relations of human hemoglobins. Proceedings of Baylor University Medical Center. 2006;**19**(3):239-245. DOI: 10.1080/08998280.2006.11928171

[9] Collins FS, Weissman SM. The molecular genetics of human hemoglobin. Progress in Nucleic Acid Research and Molecular Biology. 1984;**31**:315-462

[10] Stamatoyannopoulos G. Control of globin gene expression during development and erythroid differentiation. Experimental Hematology. 2005;**33**(3):259-271. DOI: 10.1016/j.exphem.2004.11.007

[11] Uda M, Galanello R, Sanna S, Lettre G, Sankaran VG, Chen W, et al. Genome-wide association study shows BCL11A associated with persistent fetal hemoglobin and amelioration of the phenotype of beta-thalassemia. Proceedings of the National Academy of Sciences of the United States of America. 2008;**105**(5):1620-1625

[12] Cantú I, Philipsen S. Flicking the switch: Adult hemoglobin expression in erythroid cells derived from cord blood and human induced pluripotent stem cells. Haematologica. 2014;**99**:1647-1649. DOI: 10.3324/haematol.2014.116483

[13] Waye JS, Chui DH. The alpha-globin gene cluster: Genetics and disorders. Clinical and Investigative Medicine. 2001;**24**:103-109

[14] Bernini LF, Harteveld CL. Alpha-thalassaemia. Baillière's Clinical Haematology. 1998;**11**:53-90

[15] Al-Mufti R, Hambley H, Albaiges G, Lees C, Nicolaides KH. Increased fetal erythroblasts in women who subsequently develop pre-eclampsia. Human Reproduction. 2000;**15**(7):1624-1628. DOI: 10.1093/humrep/15.7.1624

[16] Forget BG. Molecular mechanisms of beta thalassemia. In: Steinberg MH, Forget BG, Higgs DR, Nagel RL, editors. Disorders of Hemoglobin: Genetics,

Pathophysiology, and Clinical Management. Cambridge, UK: Cambridge University Press; 2001. pp. 252-276

[17] Ribeiro DM, Sonati MF. Regulation of human alpha-globin gene expression and alpha-thalassemia. Genetics and Molecular Research. 2008;7:1045-1053

[18] Gallienne AE, Dréau HM, Schuh A, Old JM, Henderson S. Ten novel mutations in the erythroid transcription factor KLF1 gene associated with increased fetal hemoglobin levels in adults. Haematologica. 2012;**97**(3):340-343. DOI: 10.3324/haematol.2011.055442

[19] Tilghman SM, Tiemeier DC, Seidman JG, et al. Intervening sequence of DNA identified in the structural portion of a mouse beta-globin gene. Proceedings of the National Academy of Sciences of the United States of America. 1978;**75**:725-729

[20] Danjou F, Anni F, Perseu L, Satta S, Dessi' C, Lai ME, et al. Genetic modifiers of beta-thalassemia and clinical severity as assessed by age at first transfusion. Haematologica. 2012;**97**:xxx. DOI: 10.3324/haematol.2011.053504

[21] Goodman M, Czelusniak J, Koop B, Tagle D, Slightom J. Globins: A case study in molecular phylogeny. Cold Spring Harbor Symposia on Quantitative Biology. 1987;**52**:875-890

[22] Proudfoot NJ, Gil A, Maniatis T. The structure of the human zeta-globin gene and a closely linked, nearly identical pseudogene. Cell. 1982;**31**:553-563

[23] Tilghman SM, Curtis PJ, Tiemeier DC, Leder P, Weissmann C. The intervening sequence of a mouse beta-globin gene is transcribed within the 15S beta-globin mRNA precursor. Proceedings of the National Academy of Sciences of the United States of America. 1978;**75**:1309-1313

[24] Krainer AR, Maniatis T, Ruskin B, Green MR. Normal and mutant human beta-globin pre-mRNAs are faithfully and efficiently spliced in vitro. Cell. 1984;**36**:993-1005

[25] Rachmilewitz EA, Giardina PJ. How I treat thalassemia. Blood. 2011;**118**(13):3479-3488

[26] Mount SM. A catalogue of splice junction sequences. Nucleic Acids Research. Chapter 3 The Normal Structure and Regulation of the Globin Gene Clusters 24. 1982;**10**:459-472

[27] Padgett RA, Grabowski PJ, Konarska MM, Seiler S, Sharp PA. Splicing of messenger RNA precursors. Annual Review of Biochemistry. 1986;**55**:1119-1150

[28] Patrinos GP, Giardine B, Riemer C, et al. Improvements in the HbVar database of human hemoglobin variants and thalassemia mutations for population and sequence variation studies. Nucleic Acids Research. 2004;**32**(Database issue):D537-DD41

[29] Giardine B, Riemer C, Hefferon T, et al. PhenCode: Connecting ENCODE data with mutations and phenotype. Human Mutation. 2007;**28**:554-562

[30] Craddock CF, Vyas P, Sharpe JA, Ayyub H, Wood WG, Higgs DR. Contrasting effects of alpha and beta globin regulatory elements on chromatin structure may be related to their different chromosomal environments. The EMBO Journal. 1995;**14**:1718-1726

[31] Lander ES, Linton LM, Birren B, et al. Initial sequencing and analysis of the human genome. Nature. 2001;**409**:860-921

[32] Bird AP. CpG-rich islands and the function of DNA methylation. Nature. 1986;**321**:209-213

[33] Fischel-Ghodsian N, Nicholls RD, Higgs DR. Unusual features of CpG-rich (HTF) islands in the human α-globin complex: Association with nonfunctional pseudogenes and presence within the 3′ portion of the ζ genes. Nucleic Acids Research. 1987;**15**:9215-9225

[34] Bird A, Taggart M, Nicholls R, Higgs D. Non-methylated CpG-rich islands at the human α-globin locus: Implications for evolution of the α-globin pseudogene. The EMBO Journal. 1987;**6**:999-1004

[35] Epner E, Rifkind RA, Marks PA. Replication of alpha and beta globin DNA sequences occurs during early S phase in murine erythroleukemia cells. Proceedings of the National Academy of Sciences of the United States of America. 1981;**78**:3058-3062

[36] Goldman MA, Holmquist GP, Gray MC, Caston LA, Nag A. Replication timing of genes and middle repetitive sequences. Science. 1984;**224**:686-692

[37] Dhar V, Mager D, Iqbal A, Schildkraut CL. The co-ordinate replication of the human β-globin gene domain reflects its transcriptional activity and nuclease hypersensitivity. Molecular and Cellular Biology. 1988;**8**:4958-4965

[38] Bulger M, Bender MA, von Doorninck JH, et al. Comparative structural and functional analysis of the olfactory receptor genes flanking the human and mouse β-globin gene clusters. Proceedings of the National Academy of Sciences of the United States of America. 2000;**97**:14560-14565

[39] Feingold EA, Forget BG. The breakpoint of a large deletion causing hereditary persistence of fetal hemoglobin occurs within an erythroid DNA domain remote from the beta-globin gene cluster. Blood. 1989;**74**:2178-2186

Chapter 6

Portal Vein Thrombosis in Patients with β-Thalassemia

Ahmed Shemran Mutlaq Alwataify, Husain Naji Alshammary and Ali Hadi Mahdi

Abstract

Beta (β)-thalassemia major, a chronic inherited hematological disease, leads to chronic anemia in affected children. One of the options for treatment is splenectomy. However, this treatment involves risk of many complications, one of which is portal vein thrombosis (PVT). The risk factors include exposure of phosphatidyl-serine of abnormal red blood cells (RBCs), increased activation, aggregation and a number of platelets and nucleated RBCs after splenectomy, increased endothelial activation, decreased nitric oxide, organ dysfunction, and thrombophilia. PVT is either complete or partial obstruction and has fatal complications, especially after splenectomy and late diagnosis without effective treatment. Diagnosis is typically made with X-ray. Generally, the incidence of PVT is between 1.7% and 9.2%. Initially, it is asymptomatic; symptoms appear gradually as thrombosis progresses. The easiest way to differentiate it from other conditions is via imaging study like Doppler ultrasound. It is usually associated with prolonged prothrombin time (PT). D-dimer and alkaline phosphatase are generally high. The treatment is the same for both the acute and chronic forms and includes the treatment of causal factors, prevention of thrombus extension, and achievement of patency of the portal vein via regular RBC transfusion and antithrombotic agents.

Keywords: β-thalassemia, portal vein thrombosis (PVT), splenectomy, thrombocytosis, doppler ultrasound, anticoagulant, transfusion

1. Introduction

1.1 Thalassemia syndrome

Beta (β)-thalassemia disease is one of the common hereditary hemolytic anemias with high prevalence in the malarial belt areas including the Mediterranean, the Middle East, Transcaucasia, Africa, South and Southeast Asia, and China. It results from mutations on chromosomes 11 and 16 for cases of β-thalassemia and alpha (α)-thalassemia, respectively, with more than 150 different mutations [1, 2]. It describes all the inherited genetic abnormalities that affect the synthesis of α- or β-globin chains and consequently normal erythropoiesis and oxygen-carrying capacity of blood.

Thalassemia is an autosomal recessive disorder classified into two main categories: α- and β-thalassemia. It commonly presents as chronic hemolytic anemia [2]. β-Thalassemia major results from homozygous or compound heterozygous mutations. It involves early presentation, multi-organ complications, frequent hospitalization, and lifelong management. Its treatment depends mainly on blood transfusion. The severe form of β-thalassemia results from defects in two globulin genes and severely reduced production of β-globulin genes [1]. β-Thalassemia minor is an asymptomatic condition due to heterozygous mutations, whereas β-thalassemia intermedia involves two defective genes and is characterized by mild-to-moderate reduction in β globulin production. It is associated with absence of regular blood transfusion and iron chelation therapy, and it may lead to serious specific complications like gall and renal stones, leg ulcer, thrombophilia, and right heart failure [1].

1.2 Pathophysiology

α-Thalassemia occurs when one or more of the four α-globin genes are missing, damaged, or changed. β-Thalassemia occurs when both β-globin genes are affected [2, 3].

β-Thalassemia is characterized by the reduced synthesis or absence of the β-globin chains in the hemoglobin molecule, resulting in accumulation of unbound α-globin chains that precipitate in erythroid precursors within bone marrow and mature erythrocytes, ultimately resulting in ineffective erythropoiesis and peripheral hemolysis [3].

Anemia stimulates the production of erythropoietin with consequent intensive but ineffective expansion of the bone marrow (up to 25–30 times more than normal), which in turn may cause the typical bone deformities. Prolonged, severe anemia and increased erythropoietic drive result in hepatosplenomegaly and extramedullary erythropoiesis [3].

The molecular and pathophysiological mechanisms underlying the disease process in patients with thalassemia have substantially increased over the past decade. There are many factors that highlight the pathophysiology of β-thalassemia. These include ineffective erythropoiesis, chronic anemia/hemolysis, and iron overload, which is secondary to increased intestinal absorption in thalassemia intermedia and excessive blood transfusion in β-thalassemia major. Thromboembolic events are common in β-thalassemia intermediate in comparison to β-thalassemia major [3], as shown in **Figure 1**.

1.3 Iron toxicity

One unit of transfused blood contains 200–250 mg of elemental iron. The body has no ability to excrete such quantities of iron. Therefore, the development of iron overload in patients receiving chronic blood transfusion is inevitable. The free or unbound iron accumulates in various organs, such as the liver, heart, pancreas, pituitary, and gonads, and causes the catalysis of injurious compounds, such as free radicals, which will begin to damage cells, leading to fibrosis or organ dysfunction [3]. Iron is highly reactive and easily alternates between two states-iron III and iron II-in a method that results in loss and gain of electrons and the generation of unsafe free radicals. This can damage lipid membranes, organelles, and DNA, ultimately causing cell death and fibrosis [1]. In a healthy individual, iron is kept safe by binding to transferrin, whereas in patients with iron overload, the transferrin capacity to bind iron is exceeded within the cells and in the plasma. This results in the production of free iron within cells or plasma, damaging many tissues in the body, which can be fatal unless treated by iron chelation therapy [4], as shown in **Figure 2**.

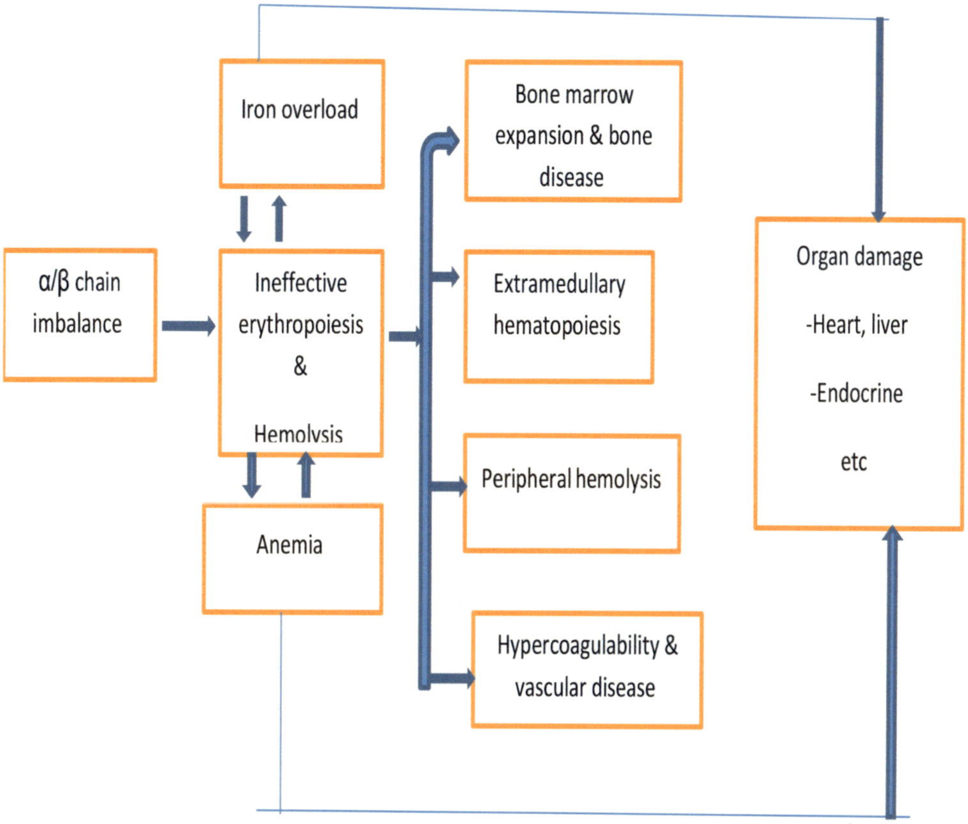

Figure 1.
Pathophysiology of patients with thalassemia.

The most common complication of β-thalassemia is iron overload, caused either by recurrent blood transfusion in β-thalassemia major or excessive iron absorption in β-thalassemia intermedia. Iron overload results in multiple organ damage [5]. Cardiac complications include cardiomyopathy, arrhythmia, and pericarditis. Liver complications include liver fibrosis and cirrhosis. Endocrine complications include hypothyroidism, pituitary failure, hypoparathyrodism, growth hormone deficiency and sex organ failure. There is also risk of embolism resulting in portal vein thrombosis and portal vein hypertension.

1.4 Treatment

Treatment involves early and regular transfusion to maintain hemoglobin (Hb) levels greater than 9–10 gm/dl. This helps to improve growth and development, reduce hepatosplenomegaly and bone deformity, improve survival, and decrease severity of disease and hemolysis (because chronic transfusion may ameliorate ineffective erythropoiesis) [1].

Chelation therapy is used to maintain safe levels of body iron at all times [5].

Dosing of the drugs involved in chelation therapy (desferrioxamine, deferasirox, jadenu, and deferiprone) and treatment require careful monitoring by serum ferritin and MRI techniques.

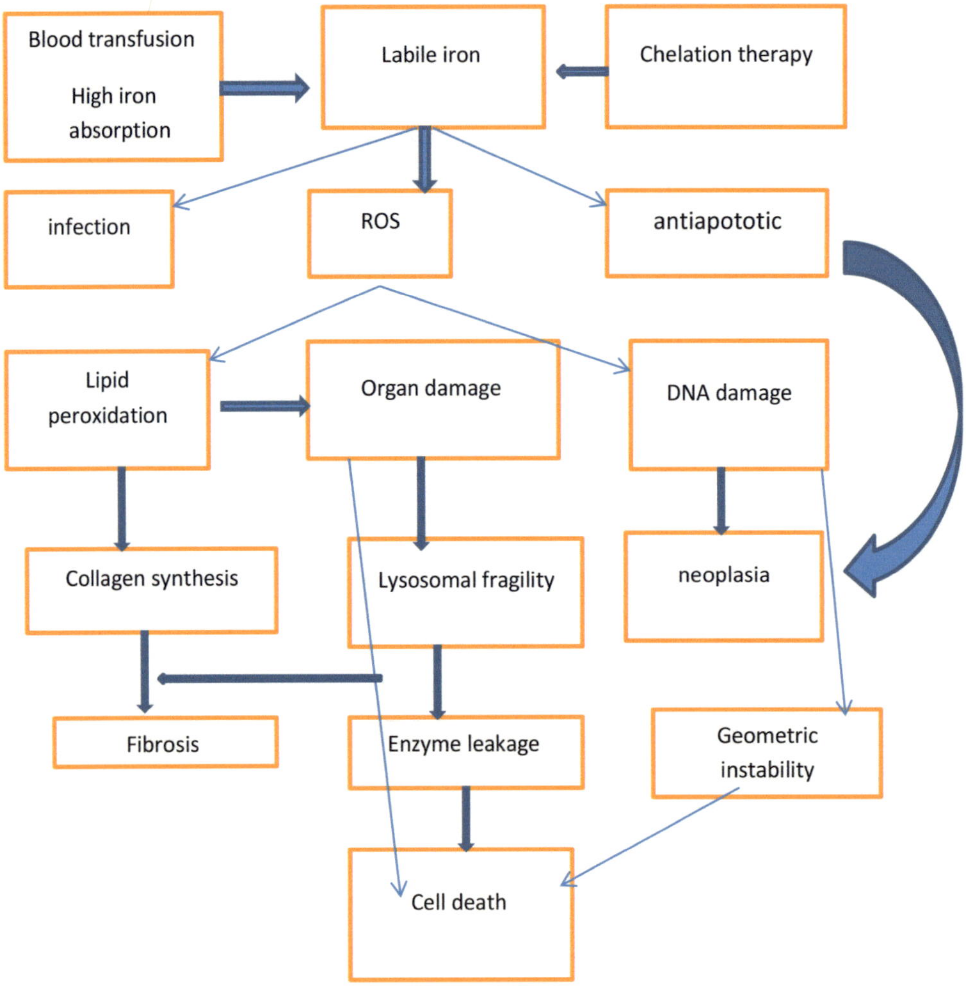

Figure 2.
Summary of mechanism of iron toxicity.

1.5 Portal vein thrombosis

PVT refers to a total or partial obstruction of the blood flow in the portal vein due to formation of thrombus or clot [6] that blocks the main portal vein going to the liver. It may result from previous use of an umbilical catheter during the neonatal period, a clotting disorder, or infection or injury. In 50% of cases, it may be idiopathic [7, 8]. It is an important cause of portal hypertension in the pediatric age group with high morbidity rate due to its main complication of upper gastrointestinal bleeding [6].

PVT is a risky disease with potentially fatal complications, mainly post splenectomy, especially if there are delays in diagnosis and treatment. It can be diagnosed early with advanced X-ray imaging [9].

Risk factors of PVT can be systemic or local [10]. Inherited systemic risk factors include:

- factor V Leiden mutation

- factor II (prothrombin) mutation
- protein C and S deficiency

Acquired systemic risk factors include:

- myeloproliferative disorder
- paroxysmal nocturnal hemoglobinuria
- hyperhomocysteinemia

Local risk factors for PVT include: focal inflammatory lesion:

- neonatal omphalitis, umbilical vein catheterization, diverticulitis, appendicitis, pancreatitis, duodenal ulcer, cholecystitis, Crohn disease, ulcerative colitis, cytomegalovirus hepatitis
- injury to the portal venous system: splenectomy, cholecystectomy, abdominal trauma, liver transplant, iatrogenic (fine needle aspiration of abdominal mass), surgical portosystemic shunting
- cirrhosis.

1.6 Clinical presentation of PVT

PVT leads to portal hypertension and can cause growth of new blood vessels called varices around the blockage. It may connect blood flow from the intestine directly to the general circulation, bypassing the liver [7].

Portal hypertension can produce an enlarged spleen, low platelet count, and gastrointestinal bleeding, and may increase production of ammonia, leading to encephalopathy [7].

1.7 Portal vein thrombosis in thalassemia

PVT is a rare serious complication post splenectomy, especially for thalassemia. It requires a very high index of suspicion to confirm early diagnosis and administer urgent therapy to prevent fatal complications such as portal vein hypertension in a thalassemia patient or bowel infarction [11].

Studies have shown that red blood cells (RBCs) from β-thalassemia major and β-thalassemia intermedia increase adhesion to endothelial cells (ECs). Also, thalassemia patients have low levels of protein C and S compared with healthy people. Prothrombin fragment 1.2 (F1.2) is a marker of thrombin generation and increases in thalassemia intermedia patients [3].

There is no role for prothrombotic mutations on the increasing incidence of coagulopathy in thalassemia patients. Studies in Lebanon and Italy show that the presence of factor V Leiden, prothrombin, and methylene tetrahydrofolate reductase mutations is not related to the increased risk of thrombosis in thalassemia patients [12].

The presence of hepatic, cardiac, or endocrine dysfunction may contribute to hypercoagulability in thalassemia; when pathologic processes overwhelm the regulatory mechanisms of hemostasis, the result is increasing amount of thrombin formation, which leads to thrombosis [12].

1.8 Incidence

The incidence of PVT in all thalassemia patients is between 1.7% and 9.2%, which is approximately ten times greater than in the normal population. The incidence is 4.4 times more prevalent in non-transfusion-dependent thalassemia patients (NTDT) than patients with transfusion-dependent thalassemia (TDT). However, PVT can occur in patients with either α- or β-thalassemia diseases [2].

Around 4% of β-thalassemia major patients and 9.6% of thalassemia intermedia patients develop TEE. The same group have shown 6 years later that 1.1% of TM patients in seven Italian centers had thrombosis [3].

PVT incidence in thalassemia is different in many centers. Some centers have reported single case reports, whereas others have reported cases in 3.3–6.6% of patients. Some centers recommend screening those patients referred for blood transfusion due to the risk of thrombosis [12].

Studies in different countries show varying incidence rates, for example, 5.5% in Al Najaf, Iraq [9], 3.12% in Babol, Iran [12], 3.5% in Ahvas, Iran [13], 3.85% in Italy [12], and 8.4% in Greece [13].

Different results could be explained by the different methods and cohorts used in different studies. Some studies included younger patients and used only Doppler ultrasound for diagnosis of PVT, while other studies used more advanced investigations in the detection of PVT, including magnetic resonance imaging (MRI), computed tomography (CT) scan, angiography, and Doppler ultrasound, in addition to using older cohorts [9].

1.9 Pathophysiology

PVT is the combination of abnormalities in different parts of the hemostatic system [2]. We discuss some of these abnormalities in the paragraphs that follow. **Figure 3** shows the pathophysiology of PVT.

Exposure of the external membrane of abnormal RBCs to phosphatidylserine (PS) results in reduction of normal dissemination of RBC membrane phospholipids. In addition, free iron stimulates lipid oxidation and increases the level of membrane-bounded hemichromes and immunoglobulin, leading to changes in the structures of spectrin and band 3 protein of RBC membrane. This results in aggregation and adhesion of abnormal RBCs to endothelial cells [2, 3].

In PVT, there is an elevated number of circulating and aggregated platelets, which are found mostly post splenectomy. The lifespan of these platelets is usually short, and they have a good response to many agonists like adenosine diphosphate (ADP), epinephrine, and collagen. PVT patients also have greater levels of plasma beta-thromboglobulin than the platelets in normal populations [3].

Increased activation of endothelial cells (due to the activation of granulocytes and monocytes) causes endothelial injury and excess level of endothelial adhesion. This leads to thromboembolic events [5].

Low nitric oxide (NO) (due to hemolysis secondary to reduced arginine level) results in pulmonary vasoconstriction and subsequently leads to chronic pulmonary thromboembolism [2].

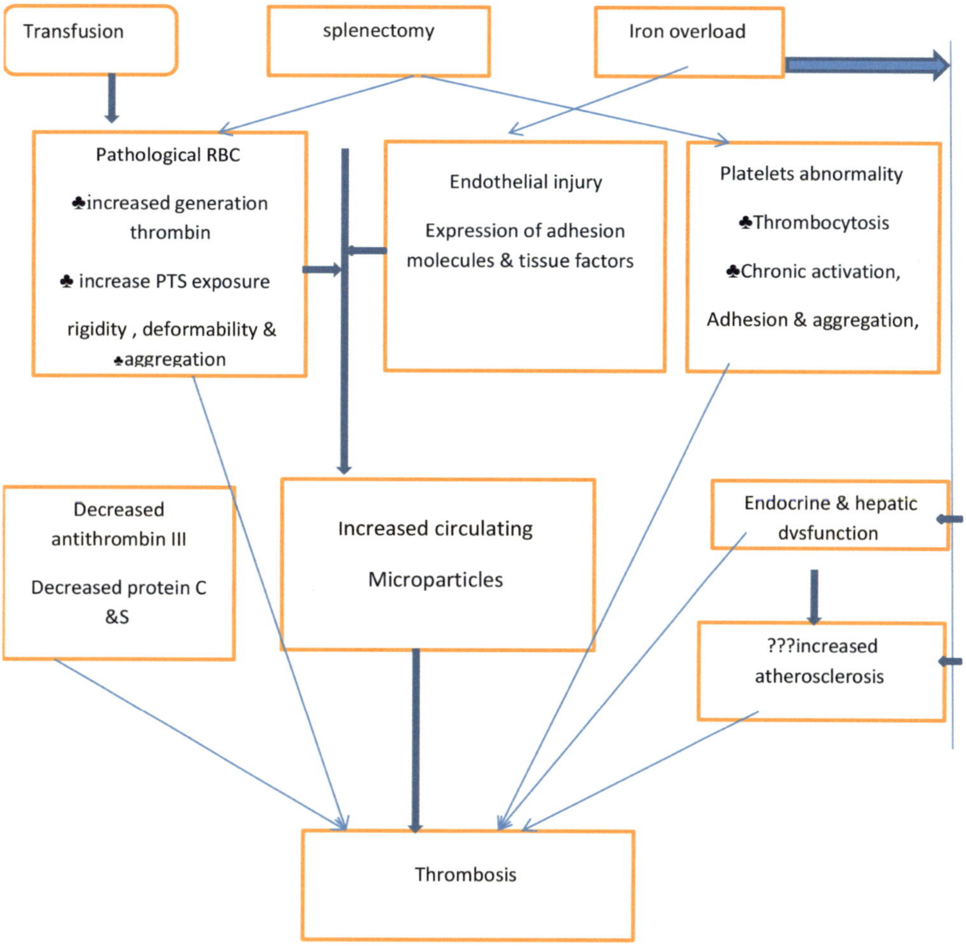

Figure 3.
Factors contributing to a hypercoagulable state and subsequent thrombotic events in thalassemia.

Thrombocytosis and nucleated RBCs (NRBCs) can occur post splenectomy, especially when the platelet count is more than 600,000/mm^3 and the NRBC count is greater than 300/mm^3 [2].

In PVT, iron overload organ failure may develop, such as cardiac siderosis, which is complicated by cardiac arrhythmia and cardiomyopathy [2].

Deficiencies of proteins C and S and elevated anti-phospholipid antibodies (i.e., lupus anticoagulant, anticardiolipin, and anti-beta 2-glycoprotein) are seen in thrombophilia [2].

Excess RBCs with negatively charged phospholipids in combination with increased cohesiveness and aggregation of RBC results in thrombus formation [2].

Prothrombin mutation is also seen in PVT [2].

1.10 Presentation

PVT presents clinically as acute or chronic. The acute form usually appears within 60 days from hospital investigation and assessment. It may present initially as upper gastrointestinal bleeding or bowel ischemia, which is suggested by an increase of bleeding, abdominal pain, abdominal distention, vomiting, and melena [10].

Initially, PVT may be asymptomatic, but symptoms gradually increase as thrombosis progresses. The easiest way to differentiate it is by using imaging study like Doppler ultrasound (to look for the presence or the absence in significant portal collateral) or by using CT of abdomen, angiograph, MRI technique [9].

1.11 Acute portal vein thrombosis

Acute PVT is associated with immediate thrombus formation, leading to partial or total obstruction of the portal vein. The cute form may present with an increase in body temperature, abdomen pain and distention, diarrhea, vomiting, rectal bleeding, and anorexia [9, 10].

1.12 Physical examination

Upon physical examination, the abdomen may be distended, guarding against internal abdominal inflammation, intestinal infarction, and perforation [9, 10].

Splenomegaly is present in 37% of patients. Sepsis may be associated with perforation, which can lead to peritonitis, shock, and even death. Ascites are rare but may develop if there is collateral circulation (mild ascites occur because of congestion of the intestinal venous without liver cirrhosis).

1.13 Chronic portal vein thrombosis

Chronic PVT is usually nearly asymptomatic except if it causes development of varices and ascites. Typically, those patients with advanced thrombosis do not remember a previous event or disease [9, 10]. Chronic PVT usually presents in the first and second decades of life as left upper abdominal pain (due to splenomegaly or splenic infraction) and recurrent gastrointestinal bleeding (in 20–40% of cases), which usually occurs in association with liver congestion or swell.

1.14 Physical examination

Splenomegaly is present in 10% of patients, but most cases present with multiple other signs. Ascites develop in 20% of cases and encephalopathy is unusual and transient. Cholangitis, obstructive jaundice, and sometimes gallstones or extra hepatic biliary obstruction occurs in 80% of cases. In addition, hypersplenism complicated by pancytopenia is common in chronic PVT.

2. Diagnosis of PVT

2.1 Laboratory tests

The liver function test is usually normal unless it is associated with cirrhosis or extrahepatic portal vein obstruction [9].

Level of prothrombin and other factors may be low (PT is prolonged), and D-dimer and alkaline phosphatase are usually high [9].

Total serum protein is usually low, especially if albumin is decreased with prolonged PVT [9].

Liver biopsy shows atrophy and regenerative nodular hyperplasia, which is due to apoptosis and compensatory arterial vasodilation in the chronic form of PVT [14].

2.2 Other lab tests

There is derangement of proteins in the hemostatic systems of β-thalassemia patients, including increased aggregation of platelets and coagulation factors (von Willebrand factor and factor VIII) as well as low levels of proteins C and anti-thrombin [9].

Annual monitoring of thrombin-generation markers by thrombin and anti-thrombin factor and D-dimer tests is recommended post-splenectomy thalassemia patients [9].

2.3 Ultrasound

Ultrasound displays hypo echoic, hyper echoic, or isochoric within the portal vein causing obstruction either completely or partially. It is the most cost-effective imaging modality, but its specificity and sensitivity vary (80–100%) depending on the patient and the experience of the administering radiologist. Its accuracy ranges from 88 to 98% [9].

Contrast-enhanced ultrasound is superior to ultrasound in detecting the patency of the portal vein [9, 15] and it is more reliable in patients with very low portal vein velocity [9].

Endoscopic ultrasound has specificity of 93% and sensitivity of 81% and is capable of diagnosing small and non-occluded thrombi. It is also more efficient than ultrasound or CT in detecting portal invasion by tumors [9, 10].

2.4 Computed tomography and magnetic resonance imaging

CT and MRI are more accurate for the detection of thrombus extension with presence or absence of collaterally vessels (that bypass the obstruction) especially after splenectomy. PVT appears as isodense to adjacent soft tissue [9]. Following administration of intravenous (IV) iodinated contrast, PVT shows a bland thrombus, which is seen as a low-density, non-enhancing defect within the portal vein. MRI is better than Doppler ultrasound in diagnosing a partial thrombosis and obstruction of the main portal venous trunk. In addition, it detects more sufficient portosplenic collaterals and portal vessels [9, 10]. However, currently used therapeutic methods have an essential in prolong life expectancy of thalassemia patients [9].

2.5 Treatments

The main goal of management is the same in both acute and chronic PVT. Treatment depends on causal factors and is used to prevent expansion of the thrombus and achieve portal vein patency. However, in chronic thrombosis, the treatment of complications associated with portal hypertension must be considered [10].

Recently, anticoagulant therapy has become the preferred treatment to obtain portal vein recanalization or patency. Other modalities of treatment should be used if there is partial or absent PVT resolution [10].

Before administering anticoagulant treatment, conditions such as new or old thrombosis and presence of thrombophilic condition or hepatic disease must be considered [10, 16].

RBC transfusion is indicated for management of β-thalassemia patients, especially NTDT patients (who are at risk of thromboembolism or have thromboembolic events because of transfusion native patients who are more risky to develop this complication [2]. Hemoglobin should be maintained at a level greater than 9 g/dL. The aim is to correct hypercoagulopathy and protect against thrombosis [3].

3. Antithrombotic agents

3.1 Antiplatelets

Acetylsalicylic acid (2–5 mg/kg/day) is the most important treatment for the prevention and management of thromboembolism in β-thalassemia patients, especially those who have undergone splenectomy and who have thrombocytosis (platelet count greater than 500,000/mm^3) [2]. Resistance to acetyl salicylic acid has been reported in some thalassemia patients, especially after spleen removal [2].

3.2 Anticoagulants

Anticoagulants are used to reduce risk of embolism, halt clot extension, and prevent recurrence [1, 17].

3.3 Anticoagulants for acute PVT

There is no randomized controlled trial regarding the use of anticoagulants in acute PVT. After 6 months of therapy, a complete recanalization has been reported in about 50% of patients, with good results in the case mesenteric vein involvement and a low incidence of complications. In contrast, there was 10% of cases of PVT which are resistant to anticoagulants [10].

Better outcomes are achieved in acute PVT when anticoagulants are given as early as possible. Rate of recanalization is about 69% if treatment is started within the first week after diagnosis. Rate of recanalization decreases to 25% when treatment starts later (e.g., the second week after diagnosis) [10].

Anticoagulant therapy should continue for 3–6 months to complete recanalization [10].

3.4 Anticoagulants for chronic PVT

Anticoagulants are only administered in 30% of chronic PVT chases, which reflects concerns about their use in the presence of varices and coagulation dysfunction. However, the number of hemorrhage episodes in chronic PVT patients on anticoagulants has not shown to increase; however, long-term follow-up may be needed. Anticoagulants have proven effective in the prevention of new thrombotic events with a low mortality rate [10].

Anticoagulants include vitamin K antagonists (warfarin), low-molecular-weight heparin (enoxaparin), and unfractionated heparin (standard heparin). Direct oral anticoagulants include direct oral anti-activated factor X (Xa) (e.g., rivaroxaban) and direct oral antithrombin (IIa) (e.g., dabigatran). These medications are used to manage thrombosis in beta-thalassemia patients [2].

In hemoglobinopathies including thalassemia, rivaroxaban is effective without increasing the risk of bleeding or thrombosis after 18 years of age [2].

4. Unfractionated standard heparin

4.1 Mechanism of action

Unfractionated standard heparin enhances the rate by which antithrombin III neutralizes the activity of several coagulation factors including factor Xa and thrombin [17].

Average half-life when administered intravenously is 60 minutes in adults. Its half-life is dose dependent (higher dose has more duration in the circulation). It has a shorter half-life than normal in patients with thrombotic disease and a longer half-life than normal in cirrhosis or uremia patients.

Contraindications of unfractionated standard heparin include recent bleeding in the central nervous system, bleeding from an inaccessible site, malignant hypertension, bacterial endocarditis, and recent surgery.

Partial thromboplastin time (PTT) may not reflect the correct degree of anticoagulation, therefore specific heparin level should be determined (heparin level is 0.35–0.70 U/ml by anti-factor Xa assay or 0.2–0.4 U/ml by protamine sulfate assay).

Protamine sulfate can neutralize heparin immediately. Because of the rapid clearance rate of heparin, stopping the infusion is adequate treatment for most patients (1 mg of protamine sulfate can neutralize 90–110 units of heparin). In addition, heparin has rapid metabolic decay and therefore it needs only one-half of a total dose of protamine.

4.2 Enoxaparin

Enoxaparin is an effective and convenient alternative to standard heparin therapy [17]. Adult patients rarely need to have their heparin level monitored, but in pediatric patients there is more diversity of response. Monitoring is critical to ensure that a therapeutic level is achieved.

PTT cannot be used to monitor heparin levels; a specific assay should be used. Once therapeutic range is achieved, routine monitoring is not required or is required infrequently.

When enoxaparin is used for prophylaxis against thrombosis, the dose is 0.5 mg/kg/12 hr subcutaneously.

4.3 Warfarin

Warfarin is an oral anticoagulant drug that acts to reduce the functional level of vitamin K-dependent factors II, IIV, IX, and X as well as proteins S and C [17]. Warfarin will reduce the level of following factors gradually depending on the half life, factor VII is firstly as the shortest its half life, followed by factor IX and X and lastly factor II, it usually needed 4-5 days to decrease all these factors. PT is a clotting test used to monitor warfarin therapy.

The most adverse effect is hemorrhage, which is related to dose or metabolism and is treated by discontinuation of the drug along with administration of vitamin K (vitamin K given is equal to the daily warfarin dose) either orally, subcutaneously, or intravenously (not intramuscularly). The parenteral route has a much longer half-life and may overshoot the correction. Sometimes, warfarin is associated with life-threatening bleeding. When severe hemorrhage develops, 15 ml/kg of fresh frozen plasma should be given in addition to vitamin K.

5. Dabigatran (Pradaxa)

5.1 Mechanism of action

Dabigatran prevents thrombus development through direct, competitive inhibition of thrombin (thrombin can convert fibrinogen to fibrin) and can inhibit free and clot-bound thrombin and thrombin-induced platelet aggregation [18, 19].

Dabigatran can be given when converting from warfarin when international normalized ratio (INR) is less than 2.0. It can also be used when converting from parenteral anticoagulant. In this case, give dabigatran 0–2 hours before next dose of parenteral drug.

When converting from dabigatran to warfarin, starting time of warfarin should be adjusted based on creatinine clearance.

6. Rivaroxaban (Xarelto)

6.1 Mechanism of action

Rivaroxaban is a factor Xa inhibitor that blocks the active site of factor Xa as well as prothrombinase activity and clot-associated factor Xa [20]. It also inhibits thrombin generation. The drug's half-life is 6–7 hours.

Rivaroxaban can lead to prolonged PT and activated partial thromboplastin time (aPTT), but it has no effect on thrombin and antithrombin activity.

It does not require routine monitoring of coagulation, but it may need anti–factor Xa chromogenic and PT assay to quantify its plasma level. However, it does require periodic assessment of renal function.

There are limited data on the drug's use in children 1 year or age or older with moderate-to-severe renal impairment. There is also no clinical data available in pediatric patients younger than 1 years old with serum creatinine above the 97.5th percentile. This drug is not recommended for use in children younger than 6 months.

6.2 Other drugs

Fetal hemoglobin-inducing agents [3, 21] like hydroxycarbamide and decitabine also appear to reduce plasma markers of thrombin formation. Hydroxycarbamide may change hypercoagulability in many ways; it may lower phospholipid expression on the surface of RBCs and platelets, and it reduces RBC adhesion to thrombospondin, a thrombin-sensitive protein. It may also reduce white blood cell (WBC) count, especially monocytes expressing transcription factor.

Hydroxyurea is a hemoglobin F stimulating agent. It increases hemoglobin F and improves the clinical symptoms of β-thalassemia disease and reduces hypercoagulability state due to decreased exposure of phosphatidylserine on the membranes of RBCs [2, 21].

6.3 Hematopoietic stem cell transplantation (HSCT)

Hematopoietic stem cell transplantation (HSCT) can normalize abnormal hemostatic derangement in β-thalassemia patients by increasing natural anticoagulant proteins and activating platelets in the circulation and reducing microparticles and Exposure of abnormal RBCs membrane to PS (results in reduction of normal dissemination of RBC membrane phospholipids) [2].

7. Conclusion

Thrombosis in β-thalassemia patients is the result of several risk factors, a combination of which is often the cause of clinical thromboembolism. Splenectomy, anemia, and iron overload are increasingly important risk factors for hypercoagulability in β-thalassemia. An individualized approach is highly indicated to create an optimal strategy for preventing occurrence of this complication through adequate post-operative monitoring via Doppler ultrasound and blood investigations, including D-dimer and PT tests.

Acknowledgements

The authors thank all the patients, their families, all investigators and medical staff in Babylon hereditary blood disease center participated in the study.

Conflict of interest

The authors declare that they have no conflict of interest.

Funding

None.

Author details

Ahmed Shemran Mutlaq Alwataify[1*], Husain Naji Alshammary[2] and Ali Hadi Mahdi[1]

1 Babylon Faculty of Medicine, Department of Pediatrics, Babylon Maternity and Children Hospital, Babylon Hereditary Blood Disease Center, Iraq

2 Babylon Faculty of Medicine, Hematology Department, Iraq

*Address all correspondence to: dr.ahmedshemranwataify@gmail.com

References

[1] Alwataify AS, Alfatlawy SS, Alshahid YA. Pulmonary hypertension in thalassemia patients. In: Artis AS, editor. Blood-Updates on Hemodynamic and on Thalassemia. London, UK: IntechOpen. DOI: 10.5722/intechopen.101052;2022

[2] Natestrinilkul R. Thromboembolism in Beta-Thalassemia Disease. London, UK: IntechOpen. DOI: 10.5772/intechopen.89313;2019

[3] Succar J, Musallamand KM, Taher AT. Thalassemia & venous thromboembolism. Mediterranean Journal of Hematology and Infectious Diseases. 2011;**3**(42)

[4] Rund D, Rachmilewitz E. β thalassemia. New England Journal of Medicine. 2005;**353**(11):1135-1146

[5] John Porter and Vip Viprakasil, Iron overload and chelation: In Cappellini, Cohen A, Porter J, Tahir A and Viprakasi V; Guide Line for the Management of Transfusion Dependent Thalassemia, 4th ed 2021: TIF: pp 42-60

[6] Priscila Menezes Ferri, Alexandre Rodrigues, Eleonora Druve, Shinfay maximilian etc.: Portal vein thrombosis in children and adolescent; 20 years experience of a pediatric hepatology reference center. Arquivos de Gastroenterologia. 2012;**49**(1):69-76

[7] Anthony K. Portal vein thrombosis. Health Line. 2018

[8] Pietrobattiata A, Luciani M, Abraldes JG, Candusso M, et al. Extrahepatic portal vein thrombosis in children and adolescents: Influence of genetic thrombophilic disorder. World Journal of Gastroenterology. 2010;**16**(48):6123-6127

[9] Kussay M, Zwain A, Mohamed SM, Al-Hakkak, Alaa A, Al-Wadess, et al. Portal vein thrombosis post splenectomy in thalassemia patients. International Journal of Research in Pharmaceutical Science. 2019;**10**(4):3356-3368

[10] Ponziani FR, Zocco MA, Campanale C, Rinninella E, Tortora A, Di Maurizio L, et al. Portal vein thrombosis: Insight into physiopathology, diagnosis, and treatment. World Journal of Gastroenterology. 2010;**16**(2):143-155

[11] Al-Charrakh HH, Al-Aqabi AS. Extensive thrombosis of the portal venous system postsplenectomy for a patient with thalassemia intermedia. Iraqi Journal of Hematology. 2014;**3**(2):132-139

[12] Hassan MN, Tahereb GM, Ahmad T, Asghar DA, et al. Correlation of splenectomy with portal vein thrombosis in β-thalassemia major. Journal of Pakistan Medical Association. 2011;**61**(8):760-762

[13] Alexakis N, Dardamanis D, Albanopoulos K, et al. Incidence, risk factors and outcome of portal vein thrombosis after laparoscopic—Assisted splenectomy in β thalassemia patients: A prospective exploratory study. Journal of Laparoendoscopic and Advanced Surgical Techniques. 2013;**23**(2):123-128

[14] Condant B, Valla D. Nonmalignant portal vein thrombosis in adult. Nature Clinical Practice Gastroenterology & Hepatology. 2006;**3**(9):505-515

[15] Ali S, Keikhaei B, Joorabian M, et al. Retrospective study of the incidence of portal vein thrombosis after splenectomy in hematological disorders: Risk factors and clinical presentation. Blood Cells,

Molecules, and Diseases. 2019;**74**:1-4. DOI: 10.1016/i.bcmd.2018.09.005. Epub 2018 Sep

[16] Valla D-C, Condat B. Portal vein thrombosis in adults: Pathophysiology, pathogenesis, management. Journal of Hematology. 2000;**32**(490):865-871. DOI: 10.1016/so 168-8278(00)80259-7

[17] Leslie J, Raffinil, Scott P. Anticoagulant and Thrombolytic Therapy. In: Kliegman RM, Behrman RE, St. Geme JW, Stanton BF, editors. Diseases of Blood. Nelson text book 20 ed. Elsevier Saunders. 2015. pp. 1710-1711

[18] Schulman S, Reilly PA. Dabigatran etexilate future directions in anticogulant treatment. Clinical and Applied Thrombosis/Hemostasis. 2009;**15**(15):32-41

[19] Hankey GJ, Eikelboom JW. Dabigtran etexilate. Circulation. 2011;**123**:1436-1450. DOI: 10.1161/circulationAHA.110.004424

[20] Thomas TF, Ganetsky V, Spinler SA. Rivaroxaban: An oral factor Xa inhibitor. Clinical Therapeutic. 2013;**35**(1):4-24

[21] Domenica Cappellini M, Khaled MM, Marcon A, Ali T. Coagulopathy in β-thalassemia: Current understanding and future perspective. Mediterranean Journal of Hematology and Infectious Diseases. 2009. Open Journal System: ISSN 2035-3006

Chapter 7

Congenital Defects with Impaired Red Blood Cell Deformability – The Role of Next-Generation Ektacytometry

Joan-Lluis Vives Corrons and Elena Krishnevskaya

Abstract

The red blood cells (RBCs) carry oxygen from the lungs to the tissues, and for this, they must be able to deform. Accordingly, an impairment of RBC deformability is the cause of RBCs trapping and removal by the spleen and haemolysis. The most common causes for the decline in red cell deformability are the RBC membrane defects (abnormal shape or ionic transport imbalance), haemoglobinopathies (increased rigidity), or enzyme deficiencies (decreased anti-oxidant defences or ATP content). The most common cause of hereditary anaemia in childhood is hereditary spherocytosis (HS), characterised by a marked RBC deformabiity. A decreased RBC deformability has been found in hereditary haemolytic anaemias (HHAs) using the new-generation osmotic gradient ektacytometry (OGE), probably due to a combination of membrane protein defects and ionic imbalance. Therefore, OGE is currently considered the gold standard for the measurement of RBC deformability and the most useful complementary tool for the differential diagnosis of HHAs. Moreover, since several new forms of treatment are currently developed for hereditary RBC defects, the clinical interest of OGE is increasing. The aim of this chapter is to provide further information about the use of RBC deformability in clinical diagnosis and the OGE as a new challenge to decrease the frequency of undiagnosed rare anaemias.

Keywords: red cell, hereditary, anaemia, deformability, haemolysis, osmotic gradient, ektacytometry

1. Introduction

The red blood cell (RBC) lives in the bloodstream about 120 days (4 months) providing oxygen and nutrients to all the tissue cells. A large part of their passage is through capillaries with a diameter ranging from 3 to 8 μm which is less than its mean cell diameter, of about 7.5 μm. Accordingly to reach all the cells, the RBCs have to adapt their shape to the dynamically changing flow conditions especially in microcirculation. This is due to its ability to deform and pass through small capillaries and recover their initial shape [1]. The basis of this extreme deformability is its characteristic discoid shape with a biconcave profile (**Figure 1a, b**). RBC deformability significantly affects blood viscosity as its decrease elevates the blood viscosity, the flow resistance and, in turn, the blood pressure [3].

RBC deformability is the result of three cell properties: (a) shape (bi-concave discocyte) depending on the surface/volume (S/V) ratio, (b) viscosity depending on intracellular haemoglobin concentration and its physicochemical state and (c) visco-elasticity largely determined by the cytoskeleton, an actin-spectrin network that underlies the lipid leafet of the membrane [4]. This structure consists of long twisted strands of alpha and beta spectrin and actin filaments that form the inner shell of the RBC and provides the basis of cell deformability (**Figure 1c**). Spectrin is bound to the lipid bilayer of the membrane at sites containing the anion exchanger, band 3 via cytoskeletal proteins, ankyrin and adducin. Some of the transmembrane proteins (such as glycophorin A) are RBC antigens and contribute to the blood group system [5]. Due to their discoid form, the normal RBCs have an S/V ratio of about 1.56, indicating an excess of the surface (membrane) with regard to the volume. This allows RBCs to change their shape without increasing the surface and, subsequently, to be highly deformable in the bloodstream and in the spleen. The spleen is a highly

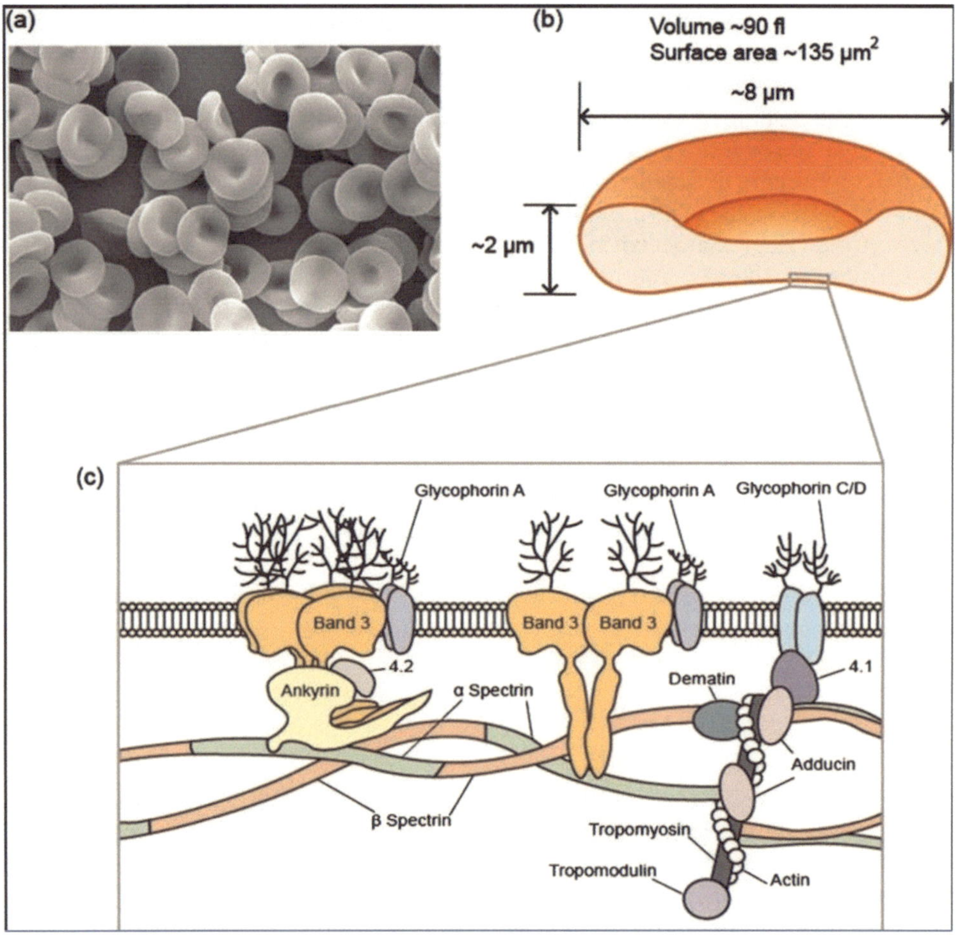

Figure 1.
Red blood cell (RBC) shape and sizes (a and b). Structure on RBC membrane skeleton (c). Source: *[2].*

DOI: http://dx.doi.org/10.5772/intechopen.109637

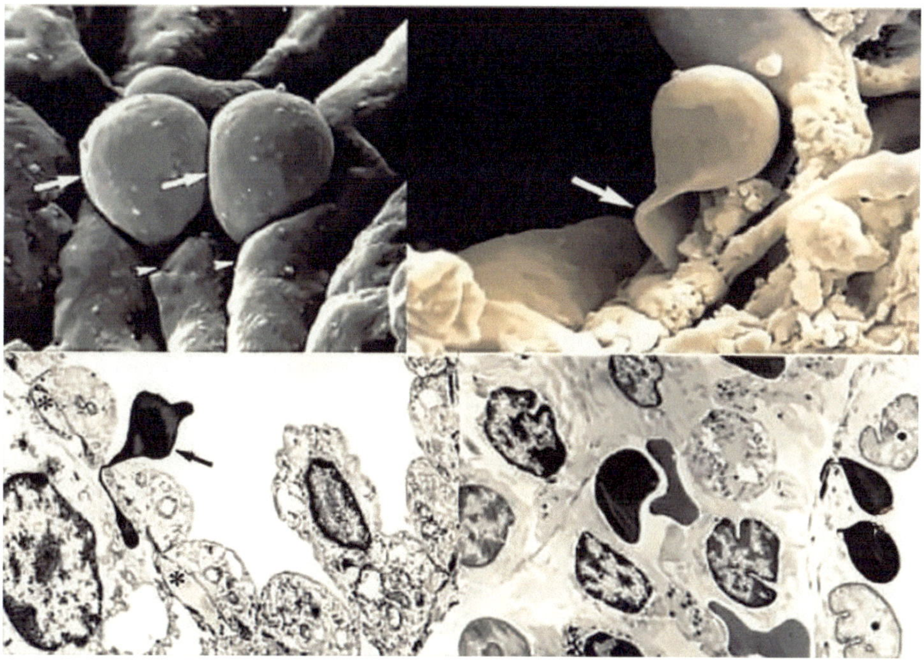

Figure 2.
Spleen red pulp observed with scanning electron microscope. RBCs trespassing the inter-endothelial slits.

vascularised organ, with a blood flow that represents around 6% of the cardiac flow. Its vasculature is formed by a complex network of capillaries and sinuses with endothelial cells that are anastomosed together without junction. Under the changes of the capillary diameter, the inter-endothelial slits (IESs) of about 0.2–0.4 μm of diameter are created, and the RBCs are obliged to pass through these IESs suffering an extreme, but reversible, deformation (**Figure 2**). Only the cells with normal deformability can overcome the IES, and due to this, this system is considered a particular splenic barrier for RBC defects [6]. When the S/V ratio for healthy RBCs is beyond the "physical fitness test" to pass through the IES, the cells are unable to adapt and are easily trapped and destroyed by the spleen [7].

2. General factors influencing RBC deformability

RBC deformability is the key physical property to ensure suitable tissue oxygenation, and thereby, it has been recognised as a sensitive indicator of RBC functionality. As shown in **Figure 1**, it depends on the structural properties of the 'horizontal' cytoskeletal components (spectrin-action-band 4.1) and the 'vertical' cytoskeletal proteins (spectrin-ankyrin-band 4.2) and their interaction with the cytosolic domain of band 3 protein (anion exchanger AE 1) or glycophorin C/D, respectively [8]. This essential deformability mechanism can, in turn, be affected by various physiological and pathological factors mainly due to intrinsic defects of cell membrane skeletal architecture [9–11], haemoglobin defects [12], mechanical damage [13] or normal RBC ageing [14–16]. There are many factors that can influence RBC deformability, but in practice, they can be summarised as follows: (a) RBC shape, (b) intracellular haemoglobin concentration (MCHC), (c) temperature, (d) osmotic pressure, (e) ATP depletion, (f) nitric oxide concentration and (g) membrane lipids and/or proteins abnormalities.

2.1 RBC shape

Due to its deformability, by shearing flow, the biconcave disk form of the RBC changes to an ellipsoid form, facilitating large reversible elastic transformation into arbitrary shapes. This enables RBCs' large deformations in the blood stream exhibiting diverse morphological features depending on physiological and physiopathological conditions [8, 14]. As mentioned before, the principal factor that makes possible the large deformations of normal RBCs is the high surface/volume (S/V) ratio, and when it changes, the abnormal deformability correlates with the pathogenesis of several RBC morphological disorders, including spherocytosis, acanthocytosis, stomatocytosis, schizocytosis and tear drop cells [15]. Since an abnormal RBC shape is the most important cause of the decreased deformability and haemolysis [16], RBC morphology examination is very helpful for the differential diagnosis of HHAs. In normal circumstances, RBC ageing is also an important cause of morphological alterations and decreased S/V ratio, as a consequence of the decrease of cellular ATP content [17].

2.2 Haemoglobin concentration

Mean cell haemoglobin concentration (MCHC) determines cytoplasmic viscosity, and it is another crucial factor determining RBC deformability. An increase of MCHC is always associated with a decrease of RBC deformability, and the two classical examples of this situation are hereditary spherocytosis (HS) and hereditary xerocytosis (HX) [1, 4, 18]. Additionally, reduced deformability in aged cells is also correlated with an increase of MCHC and RBC cytoplasmic viscosity due to cell dehydration [8].

2.3 Temperature

Waugh and Evans [19] demonstrated that the temperature plays an important role in RBC deformability. Below 25°C, deformability decreases as temperature decreases, whereas no apparent change in RBC deformability is observed between 25 and 37°C [20]. Moreover, body or febril temperature may be particularly important since the increase of body temperature decreases RBCs' elasticity and filterability. However, this has no clinical effect on the pathophysiology of haemolytic anaemia [21].

2.4 Osmotic pressure

Different osmolalities of extracellular medium can bring significant changes on RBC shape and, in turn, on its deformability. At normal physiological osmotic pressure (295 mOsm/kg H_2O), the RBCs maintain their biconcave shape and deformability, but in a hypotonic medium (< 295 mOsm/kg H_2O), they are swollen due to water intake and lyse (haemolysis). On the contrary, in a hypertonic medium (>295 mOsm/kg H_2O), the RBCs suffer a cell shrinkage and become less deformable. Although the total number of Hb molecules in RBCs, or the MCHC, does not significantly change with osmolality, the value of Hb concentration can considerably change due to water influx (Hb dilution) or efflux (Hb concentration). RBCs exhibit their maximum deformability at physiological osmotic pressure; but under either hypertonic or hypotonic conditions, their deformability decreases [22]. Interestingly, this has demonstrated that at low shear stress (1–3Pa), the RBC deformability was maximal in hypotonic conditions (225–250 mOsm/kg H_2O), which is lower than the normal plasma osmolality (290–310 mOsm/kg H_2O). This may play an important role in microcirculation processes [22].

2.5 Adenosine 5′-triphosphate (ATP) depletion

The metabolic dependence of RBC deformability has been described many years ago by Weed et al. [23], and it has been also demonstrated by techniques measuring the mechanical properties of ATP-depleted RBCs [24]. Accordingly, the ATP concentration seems to be crucial for maintaining the biconcave shape of normal RBCs, and its decrease affects the RBC shape, inducing the change from its classical biconcave shape to a flattened echinocytic shape with decreased deformability [25]. When the ATP content of RBC decreases, three factors become altered: (a) ion handling by pumps and passive transport pathways [26], (b) proteolytic activity of Ca++dependent protease calpain [27] and (c) structural integrity of the membrane architecture [28].

2.6 Nitric oxide

Nitric oxide (NO) is an important cardiovascular regulator that has an action on the vascular smooth muscle, but also as a regulatory factor in RBC deformability and aggregation [29, 30]. It has been demonstrated that the decrease of NO concentration due to the effect of NO synthase inhibitors is accompanied by a decrease of RBC deformability [31].

2.7 Disturbances of membrane lipids and/or proteins

The membrane lipids that form the double-layered surface of RBCs (the lipid bilayer) are classified as phospholipids, glycolipids and cholesterol. An increase in the cholesterol-to-phospholipid ratio (C/PL) from 1.28 to 2.0 results in a decrease in RBC filterability due to an increase of membrane rigidity [32, 33]. RBC deformability may be also affected by abnormalities of the membrane skeletal proteins such as Band 3 and glycophorin, but the most important cause of decreased RBC deformability is the existence of abnormal RBC membrane cytoskeletal proteins due to genetic defects. The best example is hereditary spherocytosis (HS), due to the decreased S/V ratio, but in homozygous hereditary elliptocytosis (HE) and pyropoikilocytosis (HPP), the deformability changes are closely related to the reduced levels of band 4.1 protein. This protein participates in the maintenance of normal membrane skeletal equilibrium and shape [34].

Haemoglobinopathies such as sickle cell disease (SCD) and thalassaemia can also alter the RBC deformability and decrease the RBC life span. SCD is an autosomal recessive inherited blood disorder due to a point mutation in β-globin gene that results in the production of HbS that under deoxygenated conditions becomes self-assembled and grows to fibres inside RBCs up to a few micrometre lengths. Due to these highly stiff HbS fibres, RBCs become elongated and sickled (sickle cells) with a significantly increase of rigidity and a decrease of deformability (**Figure 3**). After repeated

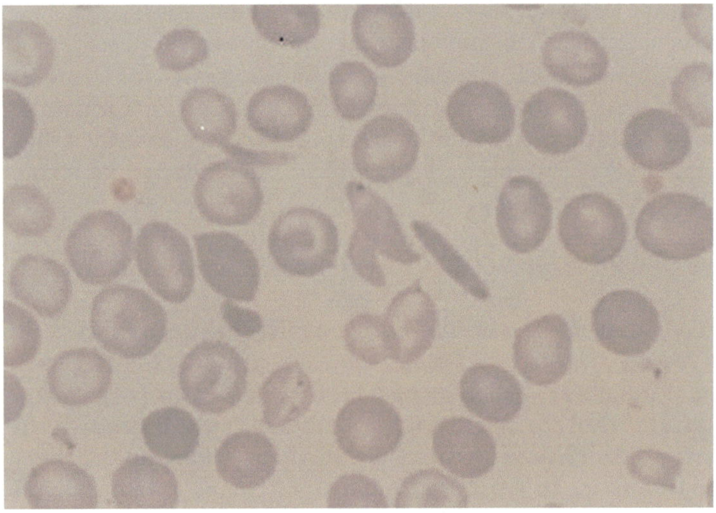

Figure 3.
Classical sickle-cell observed in MGG stained blood smear of a patient with sickle-cell disease (SCD).

sicklings, a fraction of RBCs become irreversible sickle cells (ISC), with a static rigidity that is strongly affected by the haemoglobin concentration [35]. ISCs exhibit the highest loss in deformability, and they are trapped by the spleen and retained in the microcirculation leading to severe painful vaso-occlusive crises (VOCs). Thalassaemia is characterised by a partial or total absence of one of the globin chains (**α**, **β**, **γ** or **δ**), and when this is associated with an excess of uncoupled free globin chains, these precipitate leading to RBCs' inclusions composed of denatured Hb called Heinz bodies (**Figure 4**, arrow). The limited synthesis of the globin chain, and Heinz bodies formation, may result into a local rigidification of RBC membrane [12].

3. Measurement of RBC deformability

The diagnostic procedures for the etiological investigation of HHAs have experienced a great progress in recent years [2]. The easy accessibility of RBCs from a single blood sample and the improvement of the methods used for their study have contributed to bring together the most recent knowledge in disciplines as diverse as optical and ultrastructural morphology, enzymology, metabolomics, proteomics, membrane ionic transport and, more recently, rheology or the behaviour of RBCs in suspension [36, 37]. However, the morphological examination of MGG stained blood smear still remains the simplest procedure and, in many cases, the most effective procedure for the diagnosis of HHA [1]. In the general laboratory, the classical RBC morphology examination is complemented by other simple laboratory tests such as the RBC osmotic fragility test (OFT) or the RBCs vital staining procedures with new methylene blue (NMB) for the reticulocyte count and the brilliant cresyl blue (ACB) for Heinz-bodies examination. Moreover, the incubation of a blood sample with oxidising agents, such as acetylphenylhydrazine (APH), is a very simple procedure to assess the RBC oxidant-reducing capacity. In this context, the implementation of the measure of RBC deformability may become another relatively simple diagnostic haemolysis tests. Unfortunately, the use of RBCs' deformability has not yet been

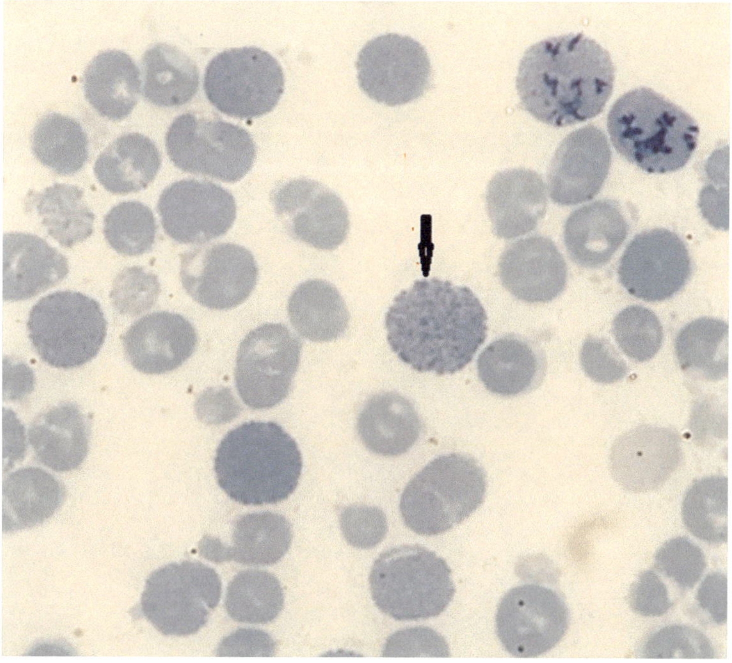

Figure 4.
Intracellular RBC hemoglobin precipitates (Heinz bodies) in a patient with HbH disease.

implemented in the diagnostic laboratories due to the several reasons that we describe here. One of these reasons is the different procedures that can be used currently to measure the deformability in both, individual or multiple RBCs.

3.1 Techniques for the measurement of individual cells

The most used procedures for measuring individual RBC deformability are based on direct measurements of single cells. This includes micropipette aspiration technique (MAT), atomic force microscopy (AFM), optical tweezers (OT) and quantitative phase imaging (QPI). A further very simple method to measure individual RBC deformability is the readout of RBC filterability after the passage of RBCs through cellulose columns [38].

a. **The micropipette aspiration technique (MAT)** was developed in the early 1980s, and since then it has been extensively used to measure the mechanical properties of RBC membranes. The measurement system consists of a micropipette, a manometer system that controls aspiration pressure and a chamber on a microscope stage from which erythrocytes are aspirated into the micropipette (**Figure 5**). By applying negative pressure, the RBC is aspirated into the micropipette, and the corresponding convexity (L) is visible in the capillary allowing measurement of the membrane elasticity [39].

b. **The atomic force microscopy (AFM)** allows viewing high-resolution topographies of materials at the atomic or molecular scale when a sharp-probe mounted at the end of a flexible cantilever deflects when interacting with the surface of a sample [40, 41]. This is detected by photodetectors, which are

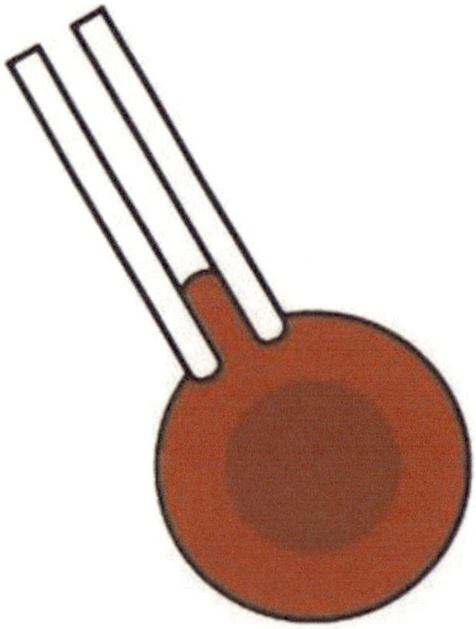

Figure 5.
The micropipette aspiration technique to measure individual RBC deformability. Source: *[2].*

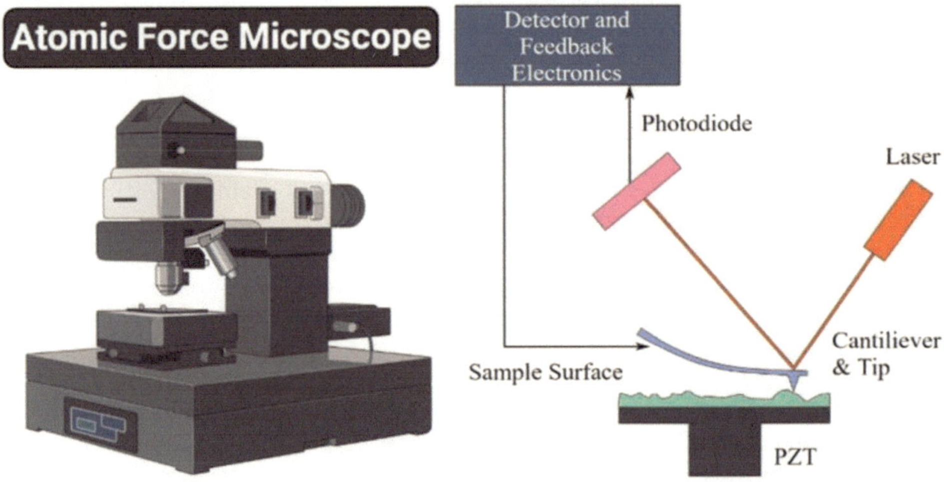

Figure 6.
Using a laser beam and a photodiode, the reflected light generated by scanning a small cantilever over the surface of the sample is collected and the images processed. Source: *S Faith Mokobi. Atomic Force Microscope (AFM)-Definition, Principle, Parts, Uses. Sagar Aryal and Wikipedia. Created with biorender.com.*

associated with the position of a laser beam reflected from the tip and can provide three-dimensional topographical images and their local mechanical properties that can be quantitatively determined from force vs. distance curves (**Figure 6**).

c. **Optical tweezers** (originally called single-beam gradient force trap) use a highly focused laser beam to hold and move the microscopic and sub-microscopic objects (nanoparticles and droplets), in a manner similar to tweezers. Optical tweezers are capable of manipulating nanometre and micron-sized dielectric particles by exerting extremely small forces via a highly focused laser beam [42]. RBC deformability can be measured by sending this highly focused laser bean through a microscope objective, and the light refraction induces trapping forces that comprise light scattering and gradient forces caused by the interaction of the light and the RBCs. The trapping forces can be determined by measuring the refractive indexes of the trapped particles and the ambient medium, laser power and particle size. When trapped particles are much smaller than the laser wavelength, the optical force can be determined by Rayleigh scattering theory; and for larger particles, such as RBCs, the Mie scattering theory is applied [43].

d. **Quantitative phase imaging (QPI)** is an optical microscopy technique in which the optical field, consisting of amplitude and phase information, is measured [44]. Since optical phase information is quantitatively related to the physical and chemical properties of a sample, QPI enables to directly analyse RBCs and quantify its size, morphology, cellular behaviour and viscoelasticity. One QPI technique used to investigate RBC deformability is the diffraction phase microscopy (DPM), and more recently, the combination of DPM analysis and mathematical modelling has been used to determine the mechanical properties of individual RBCs, such as shear modulus, bending modulus, area expansion modulus and cytoplasmic viscosity [45].

3.2 Techniques for the measurement of multiple cells

In addition to the old microfluidic approaches, some techniques, based on the measurement of RBC deformability as a function of shear stress, have become increased popularity to investigate RBC deformability and are potential tools for the routine RBC deformability measurements in clinical practice. These include the filtration method, the microfluidic filtration and the laser diffractometry.

a. **The filtration method** examines the ability of RBCs to pass through membrane filters, and it has been the first method used to measure the RBC deformability. For this, whole blood is passed through holes in a membrane filter by using the force of gravity or by applying positive or negative pressure [4, 8]. Considering the dimensions of RBCs, the pore diameters in membrane filters are 3–5 μm (e.g. Nucleopore; Corning, Acton, MA, USA). Quantification of the process is achieved either by measuring the time required to pass a certain volume of RBCs through the filter or by the pressure-flow relationship. Due to the simplicity of its components and operating principle, the filtration method has been widely used for the measurement of RBC deformability in clinical practice [46]. Unfortunately, its reproductibility is limited due to the blockage of the pores by the more rigid leukocytes or by platelet microaggregates and due to the inability to standardise the size of the pores.

b. **The microfluidic filtration** can resolve the issue of non-uniform pore size in membrane filters using a micromachining technique that produces an array of parallel microchannels.With this technique, the deformation of whole cells can be observed and measured with a microscope while they pass through the microchannels (**Figure** 7). Therefore, microfluidics filtration represents a promising, cost-effective and high-throughput method for measuring RBC deformability, with a minimum amount of blood required for the test [47–49]. The microfluidic device mimics the *in vivo* capillary blood flow system (with internal diameters measuring only a few micrometres), and RBC deformability can be measured by passing a blood sample through a funnel-shaped microconstriction. It is worth mentioning that microfluidic measurements can provide both individual RBC and population assessments of cellular deformability.

c. **Laser diffractometry** is a technique that uses light diffraction patterns produced by a laser beam traversing a sheared low haematocrit RBC suspension. When a laser beam is incident on diluted RBC suspensions, the light is scattered

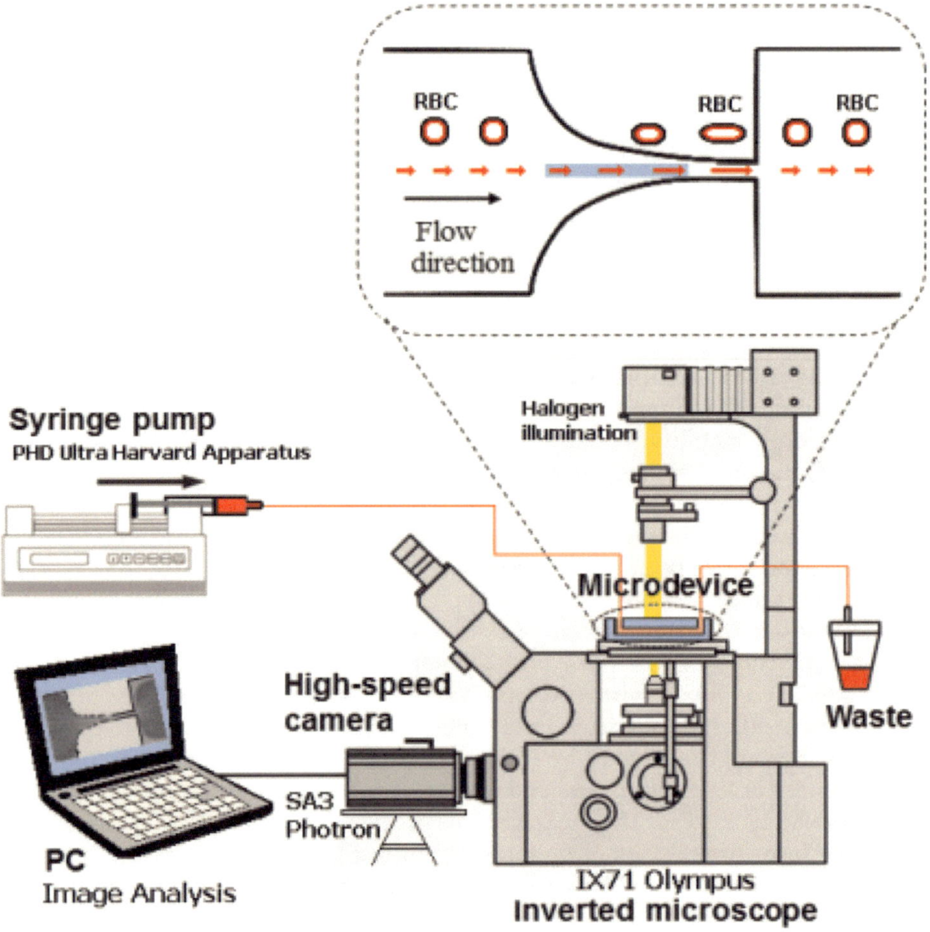

Figure 7.
Inverted microscopiy with an array of parallel microchanes to measure the RBC deformability by microfluidic filtration technique. Source: *[47].*

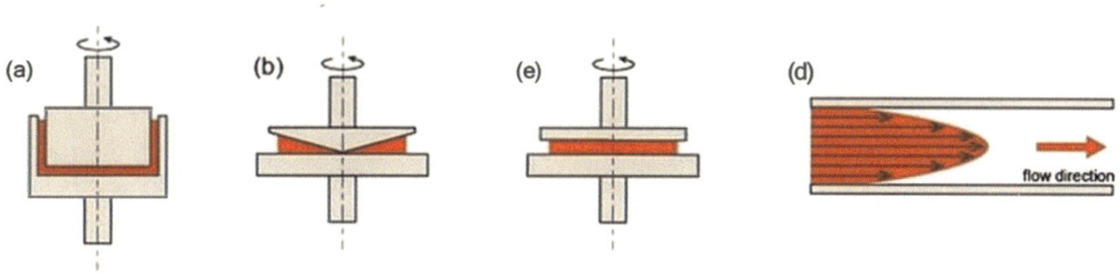

Figure 8.
Ektacytometers are using the same laser-diffraction principle but different shearing geometries: (a) concentric cylinders, (b) cone and plate, (c) parallel disks, and (d) Poiseullie slit flow.

by the RBCs population and creates a single image or diffraction pattern. The shape of the diffraction pattern reflects the average shape of hundreds or thousands of cells analysed. Due to the shape analysis of the laser diffraction pattern, laser diffractometry is also known as ektacytometry. Currently, laser diffractometry has become the primary method for testing RBC deformability, and three commercially available ektacytometers exist, using the same laser-diffraction principle but different shearing geometries [2, 50] (**Figure 8**). With these instruments, a whole blood RBC suspension in a high viscous medium is subjected to varying shear stresses that deform RBCs and different diffraction parameters are obtained. The most important diffraction parameter is the elongation index (EI) that measures the RBC deformability [50].

Due to its precision, sensitivity and convenience, laser diffractometry has become the primary method for testing RBC deformability in clinical practice and currently is represented by a new generation ektacytometer called 'Laser-assisted Optical Rotational Cell Analyser (LoRRca) MaxSis (RR Mechatronics)' (**Figure 9**). Through its Osmoscan module, which measures the RBC deformability under an osmotic gradient (OGE), the LoRRca allows to obtain a well-standardised measure of RBC deformability depending on both the shape and the position along the osmolality axis [51]. For this test, 200 μl of whole blood is needed, and four RBC parameters are defined: 1. Deformability (EImax), 2. Osmotic fragility (Omin), 3. Cellular hydration (Ohyper) and 4. Area under the curve (AUC) (**Figure 10**).

1. EImax is the maximum elongation index and represents the maximal RBC deformability at a physiological value of osmolality (Omax). EImax value is an expression of RBC membrane shape and rigidity.

2. Omin is the value of osmolality at which the EI is minimal and corresponds to the 50% lysis point as determined by the classical osmotical fragility test (OFT). Omin reflects cellular surface-to-volume ratio (S/V).

3. Ohyper is the value of osmolality in the hypertonic region that corresponds to the 50% of the EImax and reflects the cellular hydration status or intracellular viscosity. Ohyper correlates with the reciprocal function of the MCHC, and during normal RBC ageing, increased MCHC correlates with decreased RBC deformability.

4. AUC is the area under the curve (AUC) with a starting point in the hypo-osmolar region (Omin) and an ending point in the hyper-osmolar region (500 mOsm/kg).

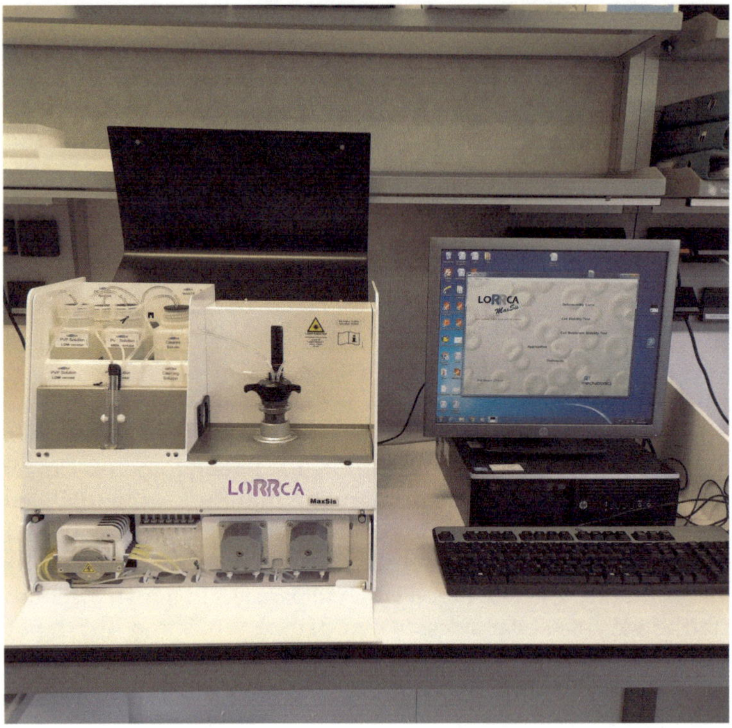

Figure 9.
Laser-assisted Optical Rotation Cell Analyzer (LoRRca MaxSis, Mechatronics, Hoorn, The Netherlands) to measure the osmotic gradient ektacytometry (OGE) parameters in hereditary hemolytic anemias.

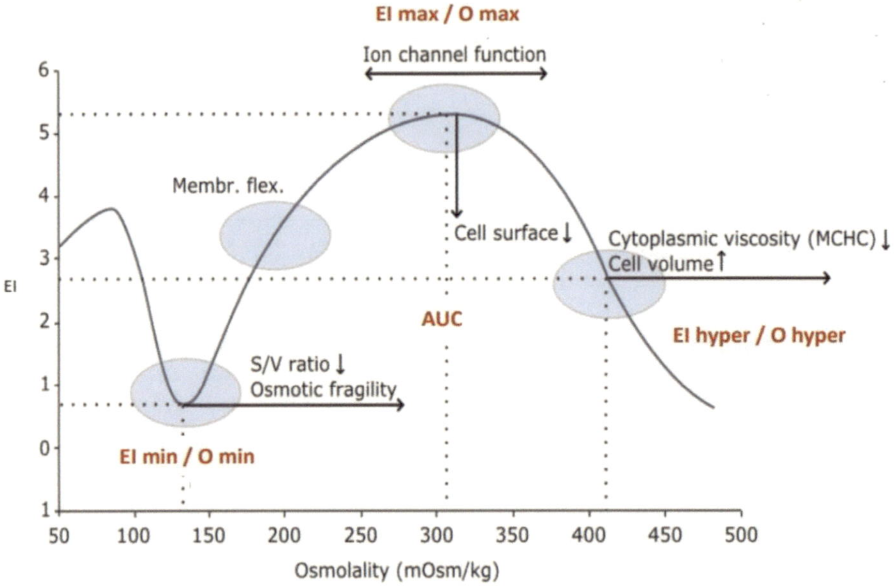

Figure 10.
Osmotic Gradient Ektacytometry (OGE) curve provides information on RBCs deformability (EI), osmotic fragility (Omin) and cell hydration (Ohyper). EI values and also the membrane rigidity are depending on both the RBCs shape and their position along the osmolality axis.

4. Congenital defects and decreased RBC deformability

The congenital defects associated with a decreased RBC deformability are an important group of rare diseases (RDs) with anaemia as their most relevant clinical manifestation. For this reason, this group of RDs are also called rare anaemias (RAs) and have been largely studied in the context of the European Network for Rare and Congenital Anemias (ENERCA). ENERCA was launched in 2002 by the European Commission (EC) to create a multidisciplinary approach for the diagnosis and clinical follow-up of patients with RAs [52]. After 2017, ENERCA has become a member of the Independent Advisory Board (IAB) of the European Reference Network (ERN) for rare haematological diseases or EuroBloodNet. Five years later, in 2022, the Thalassemia International Federation (TIF) has launched the Rare Anemias International Network (RAIN), a global community-based organisation of patient advocacy groups and industry partners which aims to advocate for the rights of people living with rare and ultra-rare anaemias worldwide. RAIN will work to raise RAs awareness through education and collaboration and to enable timely diagnosis, access to basic treatment and advanced therapies, development of specific healthcare policies and exchange best practices through more targeted and personalised services for patients with RAs (https://thalassaemia.org.cy/projects/rain/).

The most important causes of HHAs are the defects of RBC structural components: haemoglobin (haemoglobinopathies), membrane (membranopathies) and enzymes (enzymopathies). Haemoglobinopathies and enzymopathies are, in general, easily diagnosed by conventional laboratory tests such as electrophoresis, high-performance liquid chromatography (HPLC) and RBC enzyme activity measurements, respectively. On the contrary, membranopathies, despite the morphological examination of stained blood smear, allow the diagnosis in a relatively important number of cases; it is frequently hampered by several interferences. Examples of these interferences are the following: (a) the coinheritance of more than one RBC defect [18], (b) the existence of de novo mutations [53–57], (c) the overlapping of clinical variability and (d), the degree of reticulocytosis and/or to the frequent blood transfusion requirements especially in newborns and children [58–60].

According to British Committee for Standards guidelines [61], in a high percentage of cases, the consideration of patient's family history of HHA associated with typical clinical and laboratory features allows an accurate phenotypic diagnosis of RBC membranopathies. However, the recent implementation of next-generation sequencing (NGS) has drastically changed the diagnostic workflow of HHA and significantly decreased the frequency of undiagnosed cases [62–64].

From the clinical point of view, the RAs are classified into two categories: hereditary and acquired, but according to their pathophysiology, they can be classified into five groups: 1. Bone marrow (erythropoietic) defects, 2. RBC defects, 3. Iron metabolism (sideroblastic and non-sideroblastic anaemia), 4. Blood plasma discrasias (autoimmune haemolytic anaemia and related syndromes) and 5. Microcirculation diseases (haemolytic uremic syndrome and other microangiopathic disorders).

RBC deformability is affected in the RAs dealing with hereditary abnormalities of RBC components (membrane, haemoglobin or enzymes), and the acquired abnormalities are mainly due to the presence of abnormal plasma components that act on the RBC membrane (i.e. autoantibodies, crioagglutinins, plasma complement in PNH, bacteria or parasites such as plasmodium falciparum in malaria infection), to blood vessels or cardiac abnormalities (mechanic haemolysis) or to microcirculation and capillary defects (microangiopathic haemolysis). The anaemia due to haemolysis is always accompanied by a compensatory increase of bone marrow erythropoiesis and of circulating reticulocytes. If the haemolysis is fair, the increase of bone marrow erythropoiesis can maintain the haemoglobin concentration within the normal range, and there is no anaemia (compensated haemolysis). However, when bone marrow erythropoiesis is unable to compensate the intensity of the haemolysis, a typical haemolytic syndrome appears, characterised by anaemia and reticulocytosis, associated in most cases with jaundice and splenomegaly [18].

4.1 Haemoglobinopathies (structural)

These are the most frequent RBC defects when compared with membranopathies and enzymopathies and are the consequence of globin gene mutations that can alter the synthesis (thalassaemias) or the structure of haemoglobin molecule (structural haemoglobinopathies). The most frequent worldwide haemoglobinopathy is sickle cell disease (SCD), characterised by the presence of circulating sickle cells (**Figure 3**). In its homozygous form (HbSS), or combined with other haemoglobinopathies (HbSC, HbSD, HbSthal, etc.), SCD is characterised by a haemolytic syndrome of variable intensity associated with severe painful vaso-occlusive crises (VOCs) as the consequence of multiple organ micro-infarcts [65]. These VOCs are triggered by hypoxia that decreases HbS solubility, disrupts the RBC shape (sickle cells) and increases their rigidity facilitating the obliteraction of small vessels (capillaries), local intravascular haemolysis and VOC.

Haemoglobinopathies have a worldwide prevalence of about 300 million carriers, and in Europe, there are populations at risk, especially for thalassaemia, which are located in the geographical regions surrounding the Mediterranean basin (Mediterranean anaemia). HbS is not present in Caucasian individuals, but its presence in Europe is the consequence of the migration impact from people coming from Asia or African Sub-Saharan geographical regions [66]. Due to this, SCD has become one of the most important health problems in Europe and has promoted the wide implementation of neonatal screening programmes for its early detection in almost all European countries. These programmes allow to start the treatment since the first years of life, decreasing the morbidity and the mortality during early childhood [67]. Earlier studies using filtration techniques and primitive ektacytometers reported decreased deformability of sickle RBCs even under oxygenated conditions, and quantitative phase microscopy measurements demonstrated decreased membrane fluctuations on sickle RBCs [68, 69]. Recently, using membrane fluctuations, measurements of four important mechanical properties of sickle RBCs have been retrieved, and interestingly, it has been observed that in individuals with sickle cell trait (with only one abnormal allele of the Hb beta gene), their RBCs also exhibit decreased deformability when compared with healthy RBCs [70, 71].

The osmoscan curves from patients with different haemoglobinopathies are shown in **Figure 11**. They have in common a left shift of both curve tails, suggesting the existence of a different degree of RBCs dehydration depending on the type of haemoglobinopathy [72]. The most severe decrease of EImax and left shift of the osmoscan curve is observed in patients with Hb SS and HbSC, all associated with severe vase-occlusive crises. Despite the osmoscan module not considering the oxygenation of the sample, the possible deoxygenation during the analytical process may explain a partial Hb S polymerisation and, in turn, the increase of red cells dehydration and rigidity [73]. The AUC, which is an important marker of decreased

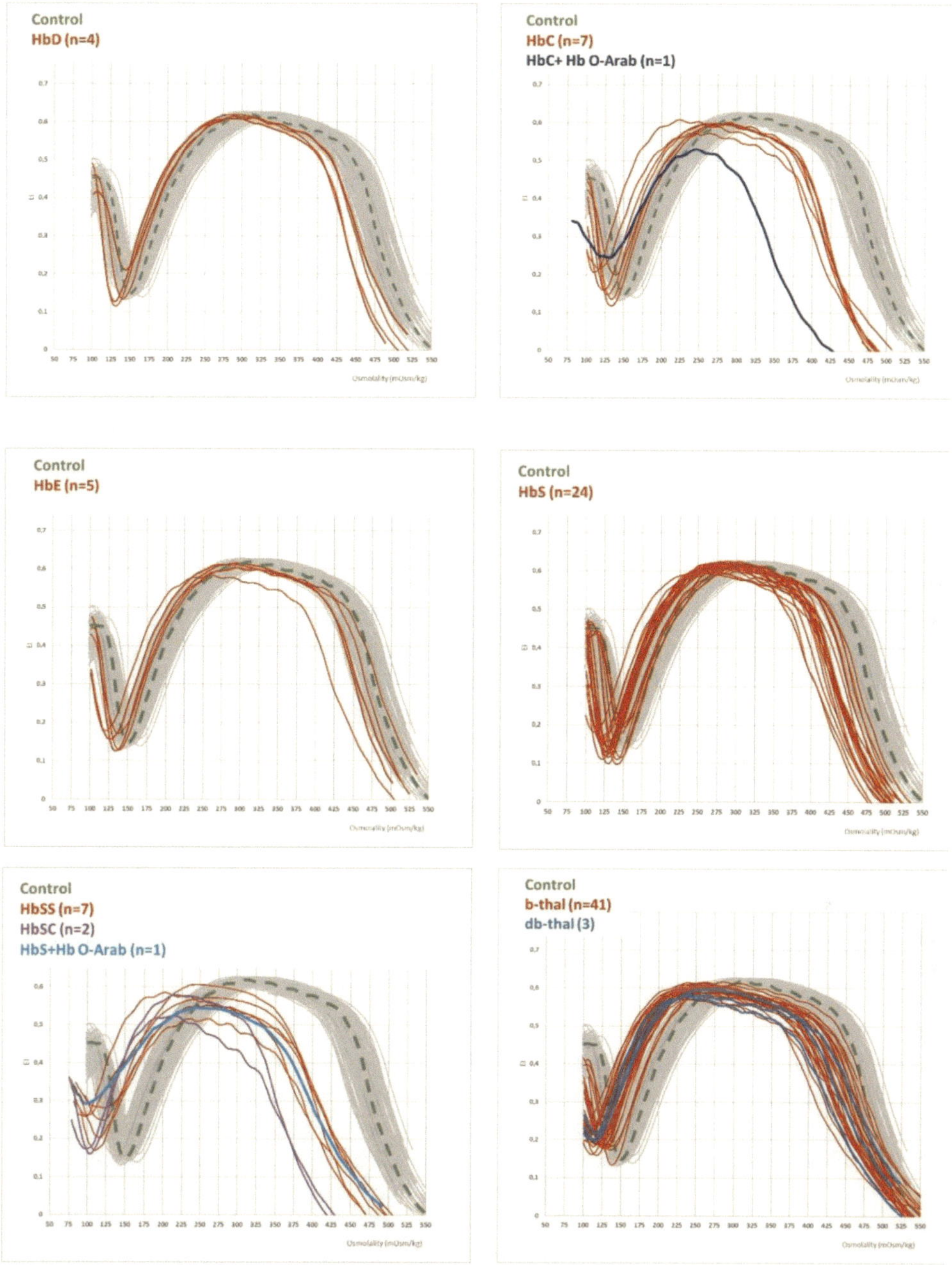

Figure 11.
Osmoscan curve profile of different hemoglobinopathies. They have in common a left shift of both curve tails, suggesting the existence of different degree of RBCs dehydration. Patients with Hb SS and HbSC, were associated with severe anemia and vase-occlusive crises.

deformability in RBC membranopathies [72], is also decreased in all the haemoglobinopathies studied by us. Carriers for Hb S, Hb C and β-thal show a similar osmoscan profile with an intermediate left shift of the curve and a less decrease of EImax (deformability) at normal osmotic value, suggesting the existence of a less degree of dehydration when compared with Hb SS and HbSC. Moreover, Hb D and Hb E show an almost normal osmoscan profile with a slight decrease of Elmax and Ohyper in accordance with their low or absent clinical expression. In addition to SCD, HbD, HbC, HbE and HbO-Arab, other structural haemoglobinopathies such as the unstable haemoglobins with intracellular haemoglobin precipitates or Heinz bodies exhibit a CNSHA of variable severity, but unlike SCD, the inheritance has an autosomal dominant pattern [72]. Interestingly, we have recently described one patient with the hyperunstable haemoglobin Bristol-Alesha, associated with severe haemolytic anaemia that exhibited the same OGE profile as β-thalassaemia [74].

4.2 Thalassaemia

Thalassaemia is the consequence of a decrease in the synthesis of a globin chain (alpha or beta) with normal Hb molecule. It is caused by the absence, decrease or defective translation of specific messenger RNA (mRNA) due to deletions or point mutations of the globin genes. While point mutations predominate in beta genes, large deletions are more frequent in alpha genes. According to the type of mutation and the severity of the decrease of globin chain, the clinical phenotype can be more or less severe [75]. In beta thalassaemia, the milder forms consist of a slight or moderate hypochromic and microcytic anaemia (thalassaemia trait), whereas the more severe clinical forms can be classified as 'thalassaemia major' or 'thalassaemia intermedia', depending on the periodicity of transfusion requirement. In alpha thalassaemia, as the genetic cluster has two genes, the mutation of a single allele, relatively common in Southern Europe, is characterised by a moderate microcytosis (MCV of about 80 fl) without anaemia (alpha thalassaemia traït), whereas if more than one allele is affected, more severe forms of alpha-thalassaemia appear like the haemoglobinopathy H (HbH) due to the formation of beta globin tetràmers (**β**4) as result of the excess or imbalance of beta chains. HbH has a similar clinical phenotype to the intermediate beta thalassaemia, but with the presence of HbH that due to its instability is sometimes undetectable. The complete loss of the four alleles (homozygous alpha-thalassaemia) is not compatible with life, leading to hydrops faetalis, abortion and death.

The differential diagnosis of thalassaemia is based on the CBC and the study of haemoglobins by electrophoresis or high-performance liquid chromatography (HPLC). In beta thalassaemia trait, there is always a characteristic increase of HbA2 fraction except in patients with concomitant iron deficiency because this condition decreases HbA2. In alpha thalassaemia trait, the haemoglobin profile is normal, and a genetic study is required for the diagnosis [75]. Concerning treatment, for the most severe cases of β-thalassaemia, it has been historically based on blood transfusions and iron chelation therapy. The only curative therapy available is allogeneic haematopoietic stem cell transplant (HSCT) from suitable donors. However, with the limited pool of donors, HSCT remains unavailable for many thalassaemic patients who may instead benefit from globin gene therapy and other modalities, which exploit recent advances in understanding of globin gene regulation [76].

RBC deformability in thalassaemia is not well known. Recently, we have demonstrated that beta-thalassaemia (**β**-thal and **δβ**-thal) shows a characteristic left shift of osmoscan curve that is different from iron deficiency anaemia [77, 78]. Probably, the decrease of one globin chain synthesis may lead to the imbalance of the α/β chains equilibrium and to the overproduction of the normal chain that may increase RBC dehydration and rigidity [8, 78, 79].

4.3 Membranopathies

Membranopathies are due to structural or functional defects of the RBC membrane proteins. In general, they are inherited as autosomal dominant pattern but transmitted with a recessive character [80–82].

Hereditary spherocytosis (HS) is the most frequent cause of HHA in Caucasians, and the most frequent proteins affected in HS are beta-spectrin (SPTB-1) Ankyrin (ANK) and Band 3 (Anion exchanger 1, AE1). Haemolysis occurs almost exclusively in the spleen, leading to splenomegaly, intermittent jaundice and cholelithiasis [57, 83]. In some patients, several complications can occur: transient erythroblastopenia crisis due to parvovirus B19 infection, severe folic acid deficiency and torpid malleolar ulcers [7, 18]. Newborns with HS and fewer can develop hazardous hyperbilirubinaemia and jaundice (neonatal icterus) associated or not with a severe anaemia. The early suspicion is essential for a prompt diagnosis and treatment, and using anticipatory guidance, adverse outcomes can be prevented [82–84]. The diagnosis of HS is based on the triad: (1) anaemia and jaundice, (2) splenomegaly and (3) spherocytosis, easily demonstrated by the peripheral blood morphological examination (**Figure 12**). The implementation of the automated haematological analysers, which perform a direct measure of the MCHC, has facilitated the use of this parameter in HS when it is increased in the presence of a high reticulocyte count. Moreover, the classical measurements of RBC osmotic fragility and criohaemolysis have been replaced by two new tests based on the measure of RBC deformability by ektacytometry and on the measure of the fluorescence intensity in RBCs after incubation with the fluorochrome, eosin-5- maleimide (EMA) by flow cytometry

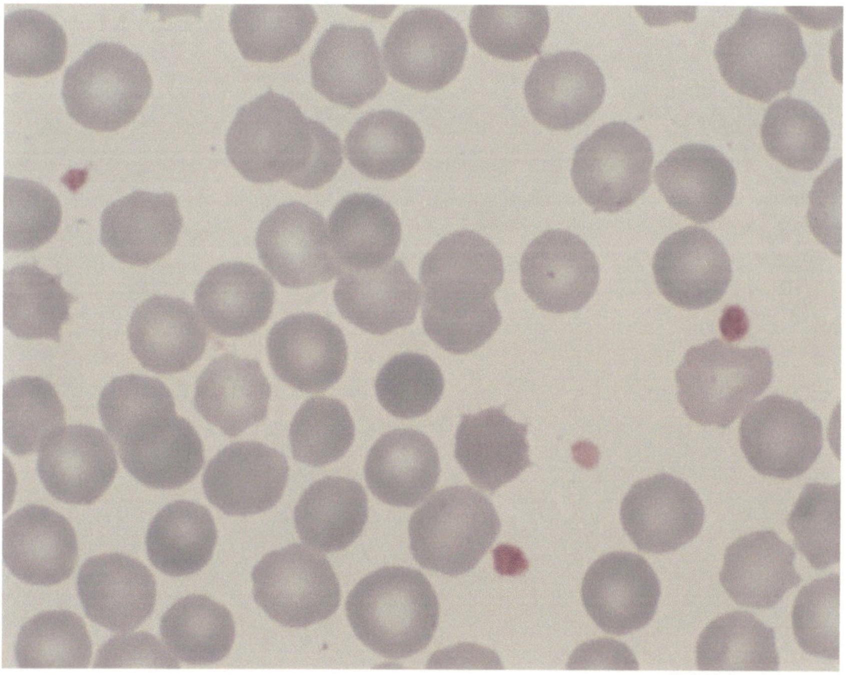

Figure 12.
Classical spherocytes observed in MGG stained blood smear from a patient with hereditary spherocytosis (HS).

(EMA-binding test). EMA binds specifically to the anion transporter (Band 3) and decreases when Band 3 decreases.

The measurement of RBC deformability using the osmoscan module of the new generation LoRRca Osmoscan from Mechatronics (**Figure 9**) has become the most sensible, accurate and reproductible method for the diagnosis of hereditary membranopathies [63]. Accordingly, when used together with the EMA-binding test, the OGE has become a reference procedure for the diagnosis of HS and an extremely useful tool for other membranopathies (**Figure 13**).

Hereditary elliptocytosis (HE) has a milder clinical expression and is characterised by the presence of more than 30% of circulating elliptocytes in peripheral blood (**Figure 14**). HE is due to a skeletal protein defect, mainly alpha-Spectrin (SPTA-1) and Band 4.1 that alters the elasticity of the membrane preventing its recovery after elongation [8, 34]. Due to this, the resulting OGE profile is characterised by a trapezoidal curve that differs from HS (**Figure 13**). However, in about 20% of patients with HE, the curve falls in the area covered by HS, making not possible to differentiate HE from HS by Osmoscan only. In the most severe clinical form of HE called hereditary pyropoikilocytosis (HPP), the SPTA-1 gene mutation in heterozygous state is associated 'in trans' with an SPTA-1 'Lely' mutation leading to severe HHA with decreased heat stability and markedly abnormal RBC morphology (**Figure 15**).

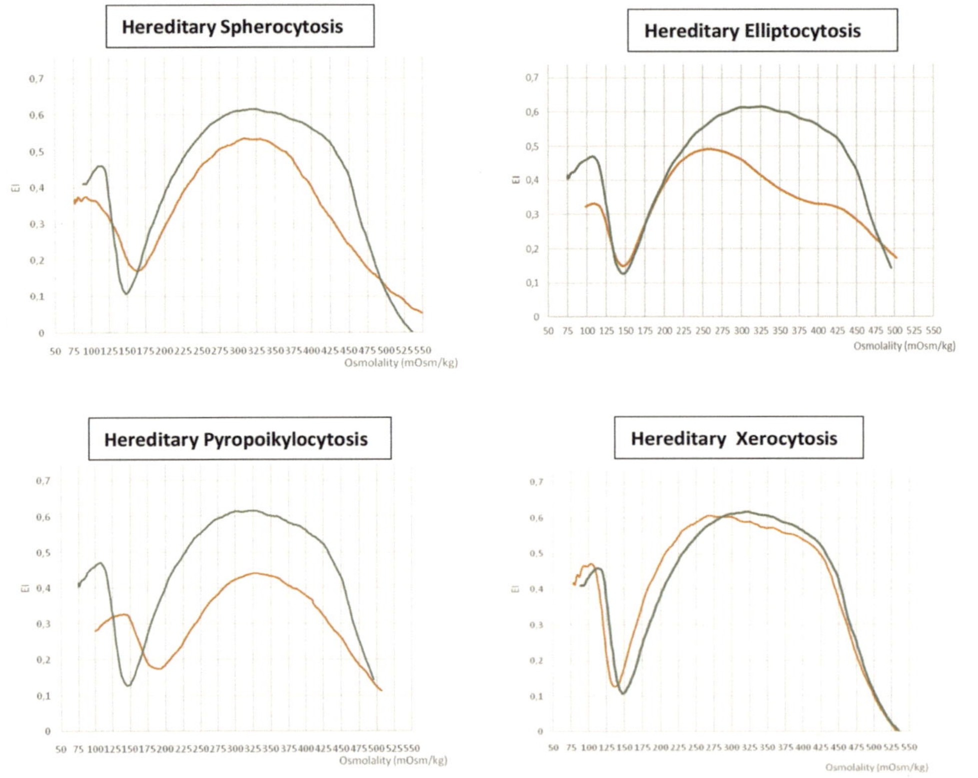

Figure 13.
Osmoscan curve profile of different membranopathies. A clear distinction between normal control (green line), hereditary spherocytosis (HS), hereditary elliptocytosis (HE), hereditary pyropoikylocytosis (HPP) and hereditary xerocytosis (HX) can be observed.

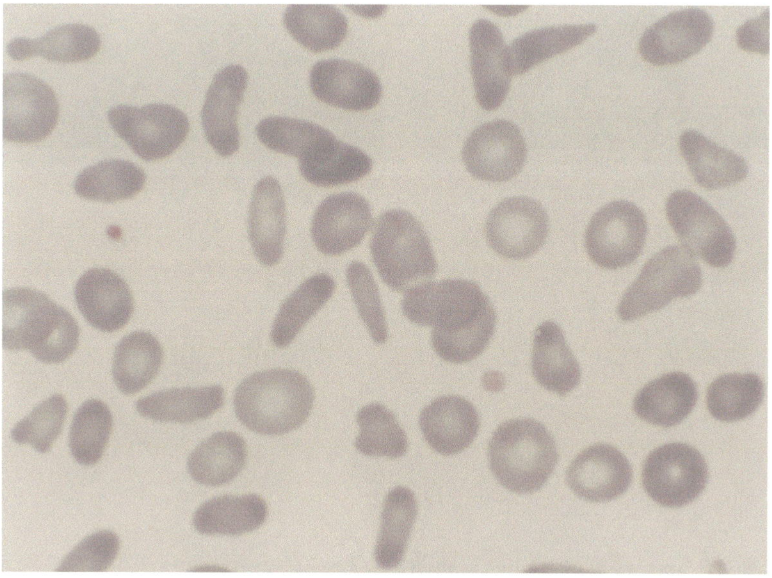

Figure 14.
Classical elliptocytes observed in MMG stained blood smear from a patient with hereditary elliptocytosis (HS).

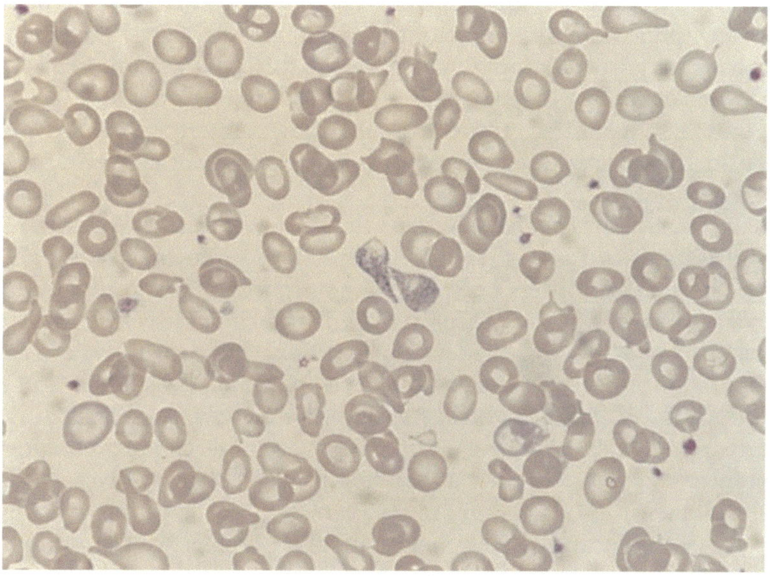

Figure 15.
Marked anisopoikilocytosis in a newborn with neonatal hemolytic anemia and jaundice due to hereditary pyropoikilocytosis (HPP).

Hereditary stomatocytosis (HSt) is an ultra-rare membranopathy where RBCs show an elongated central pallor instead of a round, and due to this they are called stomatocytes (**Figure 16**). The genuine form of HSt is the overhydrated stomatocytosis (OHSt) or hereditary hydrocytosis with chronic haemolysis and a large number of stomatocytes on peripheral blood smear. The genetic and molecular mechanism of HSt is poorly understood, but it is known that in all forms there is a disorder

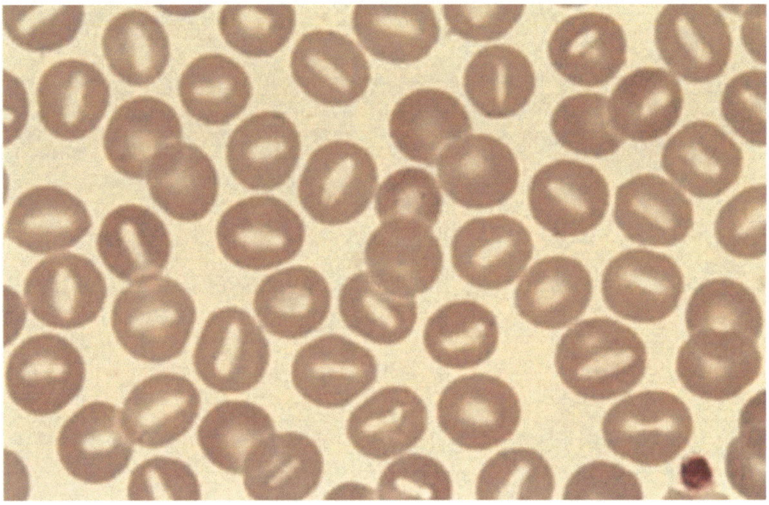

Figure 16.
Classical stomatocytes observed in MGG stained blood smear from a patient with hereditary stomatocytosis (HSt).

of the permeability to sodium and/or potassium ions associated with a markedly increased sodium permeability of about 10–40 times of normal leading to a significant increase of total mono-valent cation and water content [85, 86]. There is a variant of HSt known as cryohydrocytosis in which patient's RBCs exhibit minimal to mild changes in cation leak at physiologic temperatures, but a marked increase in monovalent cation permeability at low temperature. RBCs demonstrate a sphero-stomatocytic morphology and in some patients, heterozygous missense mutations in band 3, the anion exchanger (SLC4A1) [8, 87].

Since many years, it has been considered the existence of a second variant of HSt called dehydrated stomatocytosis (DHSt) or hereditary xerocytosis (HX). HX is the most common primary disorders of RBC ionic transport and the most clinically heterogeneous. In this disease, RBCs are dehydrated due to a cation leak, primarily of potassium, and since it is not accompanied by a proportional net gain of sodium and water, a cellular dehydration appears. Peripheral blood cell morphology is not characteristic, but few target cells and occasional erythrocytes with haemoglobin puddled to one side (eccentrocytes) can be observed [88]. When RBCs are observed in glutaraldehyde suspension, few xerocytes with the classical horse saddle shape can be seen (**Figure 17**). In HX, as in HS, the MCHC is almost always increased (34–38 g/dL), and RBC osmotic fragility decreased, reflecting cellular dehydration. OGE (LoRRca Osmoscan module) reflects a characteristic pattern of mixed reduced deformability index (decreased EImax) and dehydration (increased Ohyper) given by a leftward shift of the minimal osmolality point (**Figure 13**). The most frequent genetic mutation identified in HX affects PIEZO1 [89, 90]. but in a few HX patients, mutations in the Gardos channel, encoded by the KCNN4 gene, have been observed Clinically, HX patients with KCNN4 mutations exhibit a variable degree of anaemia associated with a higher RBC dehydration when compared with the patients with PIEZO1 gene mutations [91].

Treatment of HHA due to RBC membrane defects is always palliative, depending on the severity of anaemia. Whereas in HS and HE, splenectomy is followed by a full and partial recovery, respectively, in HX, but also in OHS, splenectomy is not recommended due to an unexplained association with thrombophilia [91].

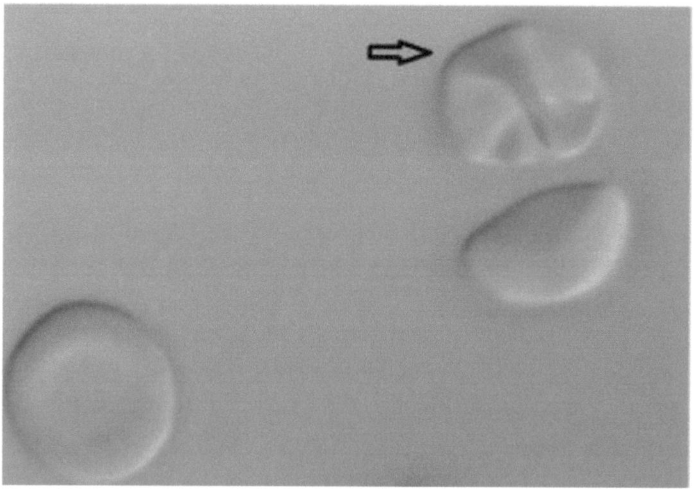

Figure 17.
RBCs in suspension with glutaraldehyde observed with optical microscopy. A xerocyte (arrow) can be observed with its classical horse ridder shape.

4.4 Erythroenzymopathies

Hereditary red blood cell (RBC) enzyme defects (erythroenzymopathies) are, in general, enzyme deficiencies, which are associated with a metabolic defect leading to CNSHA or acute haemolytic crisis with anaemia of variable severity. Some ultra-rare erythroenzymopathies are associated with neonatal cyanosis, erythrocytosis, neurological disease and myopathy.

In the circulation, RBC lifespan depends on two main metabolic pathways: 1.The anaerobic glycolysis (Embden-Meyerhof Pathway) that uses glucose to generate ATP, necessary to meet energy requirements and 2. Hexose Monophosphate Shunt (HMS) that uses NADH and NADPH to generate reduced glutathione (GSH) necessary to detoxify hydrogen peroxide (**Figure 18**). The most frequent erythroenzymopathy is glucose 6 phosphate dehydrogenase (G6PD) deficiency, followed by pyruvate kinase deficiency (PKD) and glucose-6-phosphate isomerase (GPI) deficiency. The normal RBC contains about 40 different enzymes from which 14 make up the erythrocyte metabolism. Since the mature RBCs lack mitochondria, the production of energy, in form of ATP, is entirely dependent on the anaerobic glycolysis. Accordingly, the vast majority of RBC enzyme defects described so far pertain to this metabolic pathway (**Table 1**). G6PD deficiency is in most cases asymptomatic until the patient suffers an oxidative stress induced by the ingestion of certain drugs or fava beans (favism), by infections and by other stressing clinical situations. This generates an acute haemolytic crisis with anaemia after the oxidative stress, and only few ultra-rare cases of G6PD deficiency present a life-long CNSHA as is the case of PKD and GPI. Other ultra-rare enzymopathies such as TPI, PGK and PFK exhibit a concomitant neurological impairment or myopathy, respectively.

ATP is involved in many RBC functions requiring energy, and therefore, it is essential for RBC deformability regulation [94–96]. Accordingly, RBC decreased viability in PKD and other enzymopathies pertaining to the glycolytic pathway have been suggested to depend on the decreased RBC deformability due to the decreased ATP content [8, 97]. However, the OGE profile has been found to be normal in erythroenzymopathies [98, 99] exception made of GPI deficiency where the osmoscan curve displays a significant shift to the right side and increased Ohyper.

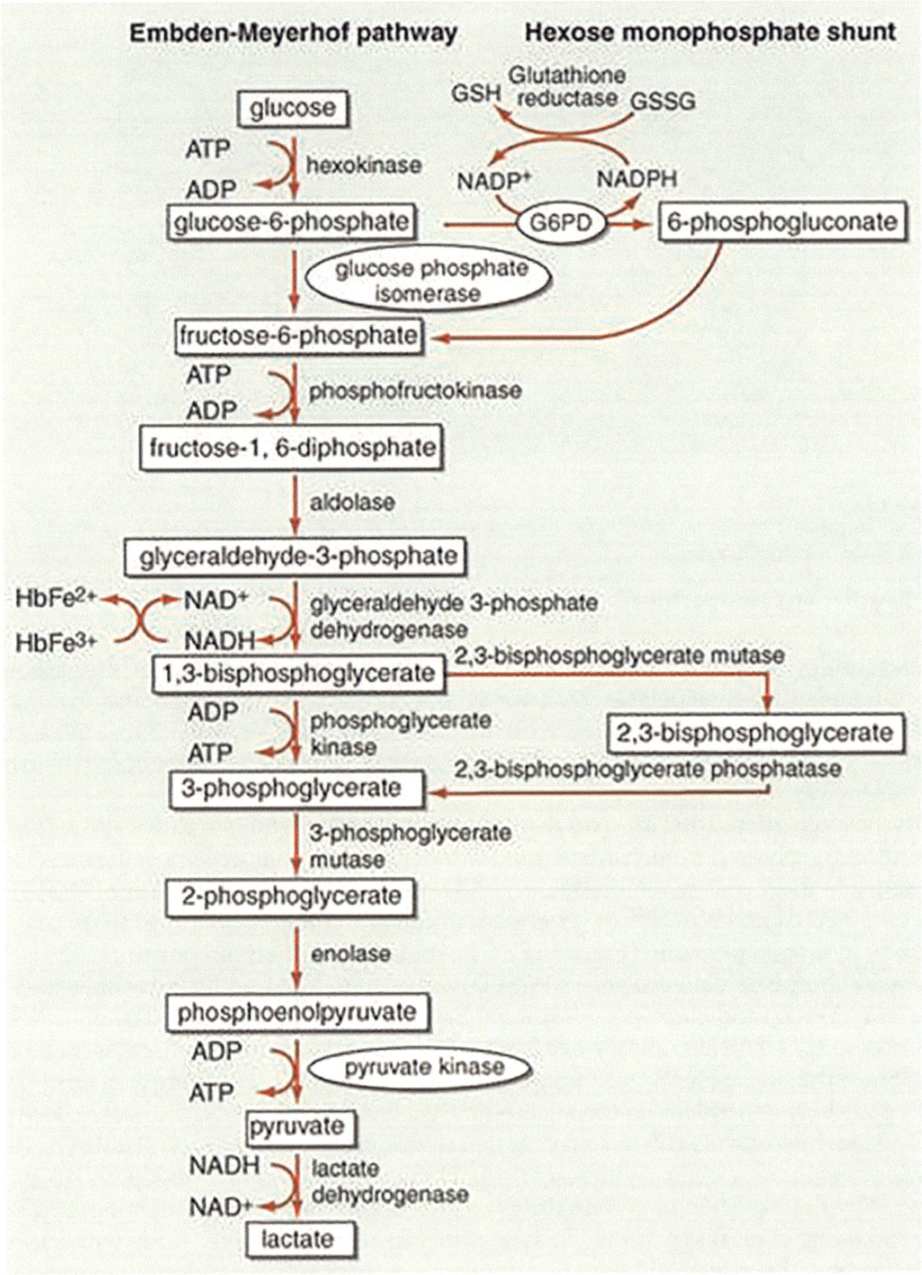

Figure 18.
RBC Metabolic pathways, Anaerobic glycolysis (Embden-Meyerhof Pathway) and the Hexose Monophosphate Shunt (HMS). Source: *[92].*

The study of the OGE parameters in 14 patients with RBC enzymopathies is shown in **Figure 19**. Three patients had asymptomatic G6PD deficiency (**Figure 19a**), five patients with homozygous PKD associated with CNSHA (**Figure 19b**), and six patients had GPI deficiency (**Figure 19c**). Interestingly, all the six cases with GPI

Enzyme	*RBC pathway*	*Clinical manifestations*	**Genetic Transmission**	**Reported cases (mutations)**
Adenosine deaminase, hyperactivity (ADA)	Nucleotide metabolism	Chronic Haemolytic Anaemia	AD	3 families (No genetic data)
Adenylate kinase (AK)	Nucleotide metabolism	• Chronic Haemolytic Anaemia	AR	12 families, (7 mutations)
Aldolase (Ald)	Glycolysis	• Chronic Haemolytic Anaemia • Neurological disease • Myopathy	AR	6 cases. (4 mutations)
Phosphofructokinase (PFK)	Glycolysis	• Chronic Haemolytic Anaemia • Myopathy	AR	50–100 cases (17 mutations)
Phosphoglycerate kinase (PGK)	Glycolysis	• Chronic Haemolytic Anaemia • Neurological disease • Myopathy	X-linked	40 cases. 19 mutations
Glucose phosphate isomerase (GPI)	Glycolysis	• Chronic Haemolytic Anaemia • Mild neurological disease	AR	>50 families (31 mutations)
Glucose-6-phosphate dehydrogenase (G6PD)	Hexose monophosphate shunt	• Acute Haemolytic Anaemia crisis triggered by oxidant drugs, • Infection and fava beans (Favism) • Chronic Haemolytic Anaemia (Very rare)	X-linked	>400 million people (>70 mutations)
6-Phosphogluconatedehydrogenase (6PGD)	Hexose monophosphate shunt	• Chronic Haemolytic Anaemia • Episodic Haemolytic Events	AR	5 cases (No genetic data)
Glutathione reductase (GR)	Glutathione metabolism	• Acute Haemolytic Anaemia crisis triggered by oxidant drugs, infection and fava beans (Favism) • Cataracts	AR	2 families, (3 mutations)
Glutathione synthetase (GS)	Glutathione metabolism	• Chronic Haemolytic Anaemia. • Neurological disease	AR	>50 families (32 mutations)
Hexokinase (HK)	Glycolysis	• Chronic Haemolytic Anaemia	AR	20 cases (5 mutations)

Enzyme	*RBC pathway*	*Clinical manifestations*	Genetic Transmission	Reported cases (mutations)
Pyrimidine-5′-nucleotidase (P5′N)	Nucleotide metabolism	• Chronic Haemolytic Anaemia	AR	>60 families. (26 mutations)
Pyruvate kinase (PK)	Glycolysis	• Chronic Haemolytic Anaemia	AR	>500 families (>200 mutations)
Triosephosphate Isomerase (TPI)	Glycolysis	• Chronic Haemolytic Anaemia • Severe Neurological Disease	AR	50–100 families (10 mutations)

***AR**: Autosomic recessive* ***AD**: Autosomic dominant.* Source: *[93].*

Table 1.
Red blood cell enzymopathies with clinical manifestations.

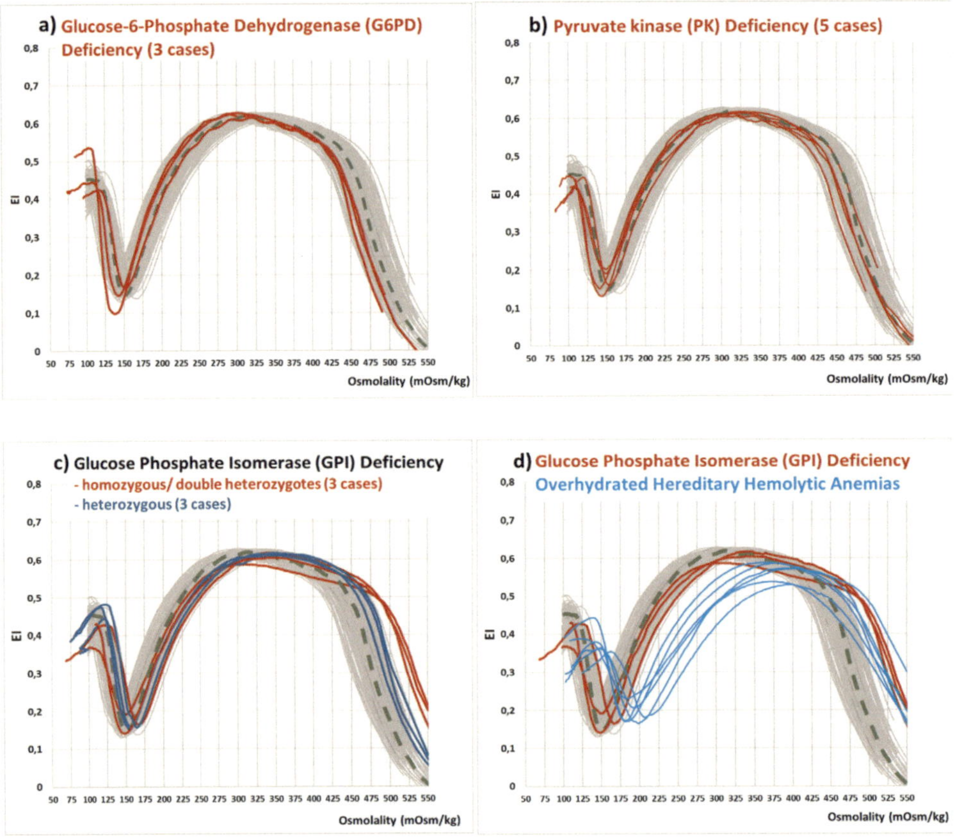

Figure 19.
Osmoscan curve profiles in three RBC enzymopathies: G6PD deficiency (a), PKD (b), GPI deficiency (c) and the comparison of the OGE of GPI deficiency with overhydrated HHA (membranopathies).

deficiency exhibit an increased Ohyper that is significantly higher in homozygous or double-heterozygous patients than in heterozygous carriers.

The finding of RBC overhydration in GPI deficiency, but not in PKD, despite both enzymes pertaining to the same glycolytic pathway is intriguing. In a previous study of a cohort of 37 patients, clinically and phenotypically diagnosed as membranopathies, we observed a particular group of six patients with a characteristic overhydrated osmoscan curve profile [100]. Three of these patients were HS type 2, according to our previous classification [63], and two patients were HHA without membranopathy phenotype but with SPTB and ANK mutations, respectively. The remaining patient was an HHA without apparent mutations and of unknown origin. In principle, our patient with GPI deficiency may be included within this group, whereas in all the patients with overhydrated HHA, the opened osmoscan profile shows an enlargement of the curve with deviation of its both sides, in GPI deficiency this enlargement affects the right side of the curve, only (**Figure 19d**). This means that in GPI deficiency, Omin and RBC osmotic fragility test (OFT) are normal but, as previously suggested [101], the membranopathies with overhydrated RBC profile are probably the associated with a concomitant unknown cannelopathy that may disturb the RBC ionic homeostasis.

In addition to haemolytic anaemia, GPI deficiency has been associated with neuromuscular dysfunction [18] because in human cells, the monomeric form of this enzyme is identical to neuroleukin (NLK), an important autocrine motility factor (AMF) AMF has an effect on the cell endoplasmic reticulum (ER) and intracellular Ca ++ homeostasis [102], and since the mature RBCs are devoid of ER, the deficient AMF/GPI protein may activate some unknown membrane ionic channel leading to overhydration. Further studies are necessary to confirm this hypothesis.

Acknowledgements

We are indebted to the European Community's Rare Diseases Research Programme that allowed to maintain the infrastructure necessary to perform this ektacytometry research project and book chapter. We are also grateful to the EU Equality/Equality Plus Projects for their partial financial support.

Compliance with ethical standards.

Conflict of interest

The authors declare that they have no conflict of interest.

Author details

Joan-Lluis Vives Corrons* and Elena Krishnevskaya
Red Blood Cell and Haematopoietic Disorders (Rare Anaemias Unit), Josep Carreras Institute for Leukaemia Research (IJC), Spain

*Address all correspondence to: jlvives@ub.edu

References

[1] Diez-Silva M, Dao M, Han J, Lim CT, Suresh S. Shape and biomechanics characteristics of human red blood cells in health and disease. MRS Bulletin. 2010;**35**:382-388. DOI: 10.1557/mrs2010.571

[2] Kim J, HoYoon L, Sehyun S. Advances in the measurement of red blood cell deformability: A brief review. Journal of Computational Biology. 2015; **1**:63-79. DOI: 10.3233/JCB-15007

[3] Banerjee R, Nageshwari K, Puniyani R. The diagnostic relevance of red cell rigidity. Clinical Hemorheology and Microcirculation. 1998;**19**:21-24

[4] Berga L, Vives Corrons JL, Feliu E, Woessner S, Rozman C Hemorheology Theoretical Basis and Clinical Applications. 1983 Salvat Editores: Barcelona

[5] Lux SE. Anatomy of the red cell membrane skeleton: Unanswered questions. Blood. 2016;**127**:187-199. DOI: 10.1182/blood-2014-12-512772 Epub 2015 Nov 4

[6] Vives Corrons JL, Berga Casafont L, Feliu FE. Concise review: How do red blood cells born, live, and die? Annals of Hematology. 2021;**100**:2425-2433. DOI: 10.1007/s00277-021-04575-z Epub 2021 Aug 3

[7] Pivkin IV, Peng Z, Karniadakis GE, Buffet PA, Dao M, Suresh S. Biomechanics of red blood cells in human spleen and consequences for physiology and disease. Proceedings of the National Academy of Sciences of the United States of America. 2016;**113**: 7804-7809. DOI: 10.1073/pnas.1606751113 Epub 2016 Jun 27. Erratum in: Proc Natl Acad Sci USA. 2017;114(22):E4521

[8] Huisjes R, Bogdanova A, van Solinge WW, Schiffelers RM, Kaestner L, van Wijk R. Squeezing for life - properties of red blood cell deformability. Frontiers in Physiology. 2018;**1**(9):656. DOI: 10.3389/fphys.2018.00656

[9] Mohandas N, Clark MR, Jacobs MS, Shohet SB. Analysis of factors regulating erythrocyte deformability. The Journal of Clinical Investigation. 1980;**66**:563-573. DOI: 10.1172/JCI109888

[10] Burton NM, Bruce LJ. Modelling the structure of the red cell membrane. Biochemistry and Cell Biology. 2011;**89**: 200-215. DOI: 10.1139/o10-154

[11] Gallagher PG. Hereditary elliptocytosis: Spectrin and protein 4.1R. Seminars in Hematology. 2004;**41**: 142-164. DOI: 10.1053/j.seminhematol.2004.01.003

[12] Reinhart WH, Sung LA. Quantitative relationship between heinz body formation and red-blood-cell deformability. Blood. 1986;**68**:1376-1383. DOI: 10.1182/blood.V68.6.1376.1376

[13] Tomaiuolo G. Biomechanical properties of red blood cells in health and disease towards microfluidics. Biomicrofluidics. 2014;**8**:051501. DOI: 10.1063/1.4895755

[14] Simmonds MJ, Meiselman HJ, Baskurt OK. Blood rheology and aging. Journal of Geriatric Cardiology. 2013;**10**: 291-301. DOI: 10.3969/j.issn.1671-5411.2013.03.010

[15] Waugh RE, Narla M, Jackson CW, Mueller TJ, Suzuki T, Dale GL. Rheologic properties of senescent erythrocytes:

Loss of surface area and volume with red blood cell age. Blood. 1992;**79**:1351-1358

[16] Bosch FH, Werre JM, Schipper L, Roerdinkholder-Stoelwinder B, Huls T, Willekens FL, et al. Determinants of red blood cell deformability in relation to cell age. European Journal of Haematology. 1994;**52**:35-41. DOI: 10.1111/ j.1600-0609.1994. tb 01282.x

[17] Meiselman HJ, Evans EA, et al. Membrane mechanical properties of ATPdepleted human erythrocytes. Blood. 1978;**52**:499-504

[18] Vives-Corrons, J. The rare anaemias. In: Wu, Z. H., editor. Rare Diseases. [Internet] London: IntechOpen; 2019. Available from: https://www. intechopen.com/chapters/69673 doi: 10.5772/intechopen.86986

[19] Waugh R, Evans EA. Thermoelasticity of red blood cell membrane. Biophysical Journal. 1989;**26**: 115-131. DOI: 10.1016/S0006-3495(79) 85239-X

[20] Singh M, Stoltz JF. Influence of temperature variation from 5 degrees C to 37 degrees C on aggregation and deformability of erythrocytes. Clinical Hemorheology and Microcirculation. 2002;**26**:1-7

[21] Williamson JR, Shanahan MO, Hochmuth RM. The influence of temperature on red cell deformability. Blood;**46**:611-624. DOI: 10.1182/blood. V46.4.611.611

[22] Heo Y, Jung H, Shin S. Osmotic deformability of erythrocytes at various shear stresses. Clinical Hemorheology and Microcirculation. 2015;**59**:211-218. DOI: 10.3233/CH-131761

[23] Weed RI, LaCelle PL. Metabolic dependence of red cell deformability. The Journal of Clinical Investigation. 1969;**48**:795-802

[24] Sheetz M, Singer S. On the mechanism of ATP-induced shape changes in human erythrocyte membranes. I. The role of the spectrin complex. Journal of Cell Biology;**73**: 638-646. DOI: 10.1083/jcb.73.3.647

[25] Gov NS, Safran SA. Red blood cell membrane fluctuations and shape controlled by ATP-induced cytoskeletal defects. Biophysical Journal. 2005;**88**: 1859-1874. DOI: 10.1529/ biophysj.104.045328 Epub 2004 Dec 21

[26] Bogdanova A, Petrushanko IY, Hernansanz-Agustín P, Martínez-Ruiz A. "Oxygen Sensing" by Na, K-ATPase: these miraculous thiols. Frontiers in Physiology. 2016;**7**:314-317. DOI: 10.3389/fphys.2016.00314

[27] Bogdanova A, Makhro A, Wang J, Lipp P, Kaestner L. Calcium in red blood cells-a perilous balance. International Journal of Molecular Sciences. 2013;**14**: 9848-9872. DOI: 10.3390/ijms14059848

[28] Gallagher PG. Update on the clinical spectrum and genetics of red blood cell membrane disorders. Current Hematology Reports. 2004;**3**:85-91 PMID: 14965483

[29] Park YK, Best CA, et al. Measurement of the nonlinear elasticity of red blood cell membranes. Physical Review. 2011;**E83**(5):051925. DOI: 10.1103/PhysRevE.83.051925

[30] Bor-Kucukatay M, Wenby RB, Meiselman HJ. Baskurt OK effects of nitric oxide on red blood cell deformability. American Journal of Physiology. Heart and Circulatory Physiology. 2003;**284**:H1577-H1584. DOI: 10.1152/ajpheart.00665.2002 Epub 2003 Jan 9

[31] Starzyk D, Korbut R, Gryglewski RJ. The role of nitric oxide in regulation of deformability of red blood cells in acute phase of endotoxaemia in rats. Journal of Physiology and Pharmacology. 1997;**48**: 731-735

[32] Cooper RA, Arner EC, Wiley JS, Shattil SJ. Modification of red cell membrane structure by cholesterol-rich lipid dispersions. A model for the primary spur cell defect. The Journal of Clinical Investigation. 1975;**55**:115-126. DOI: 10.1172/JCI107901

[33] Pfafferott C, Meiselman HJ, Hochstein P. The effect of malonyldialdehyde on erythrocyte deformability. Blood. 1982;**59**:12-15

[34] Tchernia G, Mohandas N, Shohet SB. Deficiency of skeletal membrane protein band 4.1 in homozygous hereditary elliptocytosis. Implications for erythrocyte membrane stability. The Journal of Clinical Investigation. 1981; **68**:454-460. DOI: 10.1172/jci110275

[35] Evans E, Mohandas N, et al. Static and dynamic rigidities of normal and sickle erythrocytes. Major influence of cell hemoglobin concentration. Journal of Clinical Investigation. 1984;**73**: 477-484. DOI: 10.1172/JCI111234

[36] Danielczok JG, Terriac E, Hertz L, Petkova-Kirova P, Lautenschläger F, Laschke MW, et al. Red blood cell passage of small capillaries is associated with transient Ca2+−mediated adaptations. Frontiers in Physiology. 2017;**8**:979-984. DOI: 10.3389/fphys.2017.00979

[37] Oonishi T, Sakashita K, Uyesaka N. Regulation of red blood cell filterability by Ca2+ influx and cAMP-mediated signaling pathways. The American Journal of Physiology. 1997;**273**:C1828-C1834. DOI: 10.1152/ajpcell.1997.273.6.C1828

[38] Brooks DE, Evans EA. In: Chien S, Dormandy J, Ernst E, Matrai A, editors. Rheology of blood cells, Clinical Hemorheology. Developments in Cardiovascular Medicine. Vol. 74. Dordrecht: Springer; 1987. DOI: 10.1007/978-94-009-4285-1_3

[39] Reid H, Barnes A, Lock P, Dormandy J, Dormandy T. A simple method for measuring erythrocyte deformability. Journal of Clinical Pathology. 1976;**29**:855-858. DOI: 10.1136/jcp.29.9.855

[40] Binnig G, Quate CF, Gerber C. Atomic force microscopi. Physical Review Letters. 1986, 1986;**56**:930. DOI: 10.1103/PhysRevLett.56.930

[41] Dulińska I, Targosz M, Strojny W, Lekka M, Czuba P, Balwierz W, et al. Stiffness of normal and pathological erythrocytes studied by means of atomic force microscopy. Journal of Biochemical and Biophysical Methods. 2006;**66**:1-11. DOI: 10.1016/j.jbbm.2005.11.003 Epub 2006 Jan 17

[42] Ashkin A. Acceleration and trapping of particles by radiation pressure. Physical Review Letters. 1970;**24**:156-159

[43] Ashkin A, Dziedzic JM, Bjorkholm JE, Chu S. Observation of a single-beam gradient force optical trap for dielectric particles. Optics Letters. 1986;**11**:288-290. DOI: 10.1364/OL.11.000288

[44] Popescu, Gabriel. Quantitative Phase Imaging of Cells and Tissues. 1st ed. 2011: New York: McGraw-Hill. ISBN: 9780071663427

[45] Shaked NT, Satterwhite LL, Telen MJ, Truskey GA, Wax A.

Quantitative microscopy and nanoscopy of sickle red blood cells performed by wide field digital interferometry. Journal of Biomedical Optics. 2011;**16**:030506. DOI: 10.1117/1.3556717

[46] Shevkoplyas SS, Yoshida T, Gifford SC, Bitensky MW. Direct measurement of the impact of impaired erythrocyte deformability on microvascular network perfusion in a microfluidic device. Lab on a Chip. 2006;**6**:914-920. DOI: 10.1039/b601554a Epub 2006 May 3

[47] Faustino V, Rodrigues RO, Pinho D, Costa E, Santos-Silva A, Miranda V, et al. A microfluidic deformability assessment of pathological red blood cells flowing in a hyperbolic converging microchannel. Micromachines. 2019;**10**:645-652. DOI: 10.3390/mi10100645

[48] Dobbe JG, Streekstra GJ, Hardeman MR, Ince C, Grimbergen CA. Measurement of the distribution of red blood cell deformability using an automated rheoscope. Cytometry. 2002; **50**:313-325. DOI: 10.1002/cyto.10171

[49] Shin S, Hou JX, Suh JS, Singh M. Validation and application of a microfluidic ektacytometer (RheoScan-D) in measuring erythrocyte deformability. Clinical Hemorheology and Microcirculation. 2007;**37**:319-328

[50] Baskurt OK, Hardeman MR, Uyuklu M, Ulker P, Cengiz M, Nemeth N, et al. Comparison of three commercially available ektacytometers with different shearing geometries. Biorheology. 2009;**46**:251-264. DOI: 10.3233/BIR-2009-0536

[51] Nemeth N, Kiss F, Miszti-Blasius K. Interpretation of osmotic gradient ektacytometry (osmoscan) data: A comparative study for methodological standards. Scandinavian Journal of Clinical and Laboratory Investigation. 2015;**75**:213-222. DOI: 10.3109/00365513.2014.993695 Epub 2015 Jan 16

[52] Vives Corrons JL, Mañú Pereira MM. Enerca: The European Network for Patients with Rare Anaemias. Journal of Rare Disorders: Diagnosis & Therapy. 2015;**1**:1. DOI: 10.21767/2380-7245.10009

[53] Vives Corrons JL, Mañú Pereira MM, Trujillo JP, Surrallés J, Sevilla J. Anemias raras y fallos medulares hereditarios. Arbor. 2018;**194**(789):a463. DOI: 10.3989/arbor. 2018.789n300

[54] King MJ, Zanella A. Hereditary red cell membrane disorders and laboratory diagnostic testing. International Journal of Laboratory Hematology. 2013;**35**: 237-243. DOI: 10.1111/ijlh.12070 Epub 2013 Mar 11

[55] Christensen RD, Yaish HM, Gallagher PG. A pediatrician's practical guide to diagnosing and treating hereditary spherocytosis in neonates. Pediatrics. 2015;**135**:1107-1114. DOI: 10.1542/peds.2014-3516

[56] An X, Mohandas N. Disorders of red cell membrane. British Journal of Haematology. 2008;**141**:367-375. DOI: 10.1111/j.1365-2141.2008.07091.x

[57] Narla J, Mohandas N. Red cell membrane disorders. International Journal of Laboratory Hematology. 2017; **39**(Suppl. 1):47-52. DOI: 10.1111/ijlh.12657

[58] Roy NB, Wilson EA, Henderson S, Wray K, Babbs C, Okoli S, et al. A novel 33-Gene targeted resequencing panel provides accurate, clinical-grade diagnosis and improves patient management for rare inherited anaemias. British Journal of

Haematology. 2016;**175**:318-330. DOI: 10.1111/bjh.14221 Epub 2016 Jul 19

[59] Kim Y, Park J, Kim M. Diagnostic approaches for inherited hemolytic anemia in the genetic era. Blood Research. 2017;**52**:84-94. DOI: 10.5045/br.2017.52.2.84 Epub 2017 Jun 22

[60] Iolascon A, Andolfo I, Russo R. Advances in understanding the pathogenesis of red cell membrane disorders. British Journal of Haematology. 2019;**187**:13-24. DOI: 10.1111/bjh.16126 Epub 2019 Jul 31

[61] Bolton-Maggs PH, Langer JC, Iolascon A, Tittensor P, King MJ. General haematology task force of the British committee for standards in haematology. Guidelines for the diagnosis and management of hereditary spherocytosis–2011 update. British Journal of Haematology. 2012;**156**:37-49. DOI: 10.1111/j.1365-2141.2011.08921.x Epub 2011 Nov 5

[62] Lazarova E, Gulbis B, Oirschot B, van Wijk R. Next-generation osmotic gradient ektacytometry for the diagnosis of hereditary spherocytosis: interlaboratory method validation and experience. Clinical Chemistry and Laboratory Medicine. 2017;**55**:394-402. DOI: 10.1515/cclm-2016-0290

[63] Llaudet-Planas E, Vives-Corrons JL, Rizzuto V, Gómez-Ramírez P, Sevilla Navarro J, Coll Sibina MT, et al. Osmotic gradient ektacytometry: A valuable screening test for hereditary spherocytosis and other red blood cell membrane disorders. International Journal of Laboratory Hematology. 2017; **40**:94-102. DOI: 10.1111/ijlh.12746

[64] van Vuren A, van der Zwaag B, Huisjes R, Lak N, Bierings M, Gerritsen E, et al. The complexity of genotype-phenotype correlations in hereditary spherocytosis: A cohort of 95 patients: Genotype-phenotype correlation in hereditary spherocytosis. Hema. 2019;**3**(4):e276. DOI: 10.1097/HS9.0000000000000276

[65] Piel FB, Steinberg MH, Rees DC. Sickle cell disease. The New England Journal of Medicine. 2017;**376**:1561-1573. DOI: 10.1056/NEJMra1510865

[66] Aguilar Martinez P, Angastiniotis M, Eleftheriou A, et al. Haemoglobinopathies in Europe: Health & migration policy perspectives. Orphanet Journal of Rare Diseases. 2014;**9**:97-108. DOI: 10.1186/1750-1172-9-97

[67] Daniel Y, Elion J, Allaf B, Badens C, Bouva MJ, Brincat I, et al. Newborn screening for sickle cell disease in Europe. International Journal of Neonatal Screening. 2019;**5**:15-21. DOI: 10.3390/ijns5010015

[68] Maciaszek JL. Lykotrafitis G Sickle cell trait human erythrocytes are significantly stiffer than normal. Journal of Biomechanics. 2011;**44**:657-661. DOI: 10.1016/j.jbiomech.2010.11.008 Epub 2010 Dec 8

[69] Rees DC, Williams TN, Gladwin MT. Sickle-cell disease. Lancet. 2010;**376**: 2018-2031. DOI: 10.1016/S0140-6736(10)61029-X Epub 2010 Dec 3

[70] Bunn HF. Pathogenesis and treatment of sickle cell disease. The New England Journal of Medicine. 1997;**337**: 762-369. DOI: 10.1056/NEJM199709113371107

[71] Barabino GA, Platt MO, Kaul DK. Sickle cell biomechanics. Annual Review of Biomedical Engineering. 2010;**12**: 345-367. DOI: 10.1146/annurev-bioeng-070909-105339

[72] Krishnevskaya E, Molero M, Ancochea A, Leguizamón V,

Hernandez I, Vives Corrons JL. New generation ektacytometry study of red blood cells in different hemoglobinopathies and thalassemia. Thalassemia Reports. 2023. (in press). Available from: www.mdpi.com/journal/thalassrep

[73] Gallagher PG. Disorders of erythrocyte hydration. Blood. 2017;**130**: 2699-2708. DOI: 10.1182/blood-2017-04-590810 Epub 2017 Oct 19

[74] Jl VC, Bain BJ. Haemoglobin Bristol-Alesha in a child with non-spherocytic severe haemolytic anaemia and marked anisochromic poikilocytosis with basophilic stippling and amorphous intracellular content. Blood Cells, Molecules & Diseases. 2022;**94**:102652. DOI: 10.1016/j.bcmd.2022.102652 Epub 2022 Jan 22

[75] Angastiniotis M, Thalassemias LS. An overview. International Journal of Neonatal Screening. 2019;**5**:16-21. DOI: 10.3390/ijns5010016

[76] Thompson AA, Walters MC, Kwiatkowski J, Rasko JEJ, Ribeil JA, Hongeng S, et al. Gene therapy in patients with transfusion-dependent β-Thalassemia. The New England Journal of Medicine. 2018;**378**:1479-1493. DOI: 10.1056/NEJMoa1705342

[77] Krishnevskaya E, Payán-Pernía S, Hernández-Rodríguez I, Remacha Sevilla ÁF, Ancochea Serra Á, Morales-Indiano C, et al. Distinguishing iron deficiency from beta-thalassemia trait by new generation ektacytometry. International Journal of Laboratory Hematology. 2021; **43**:e58-e60. DOI: 10.1111/ijlh.13362 Epub 2020 Oct 22

[78] Méndez-Mora L, Cabello-Fusarés M, Ferré-Torres J, Riera-Llobet C, Krishnevskaya E, Trejo-Soto C, et al. Blood rheological characterization of β-thalassemia trait and iron deficiency anemia using front microrheometry. Frontiers in Physiology. 2021;**21-12**: 761411. DOI: 10.3389/fphys.2021.761411

[79] Clark MR, Mohandas N, Shohet SB. Osmotic gradient ektacytometry: Comprehensive characterization of red cell volume and surface maintenance. Blood. 1983;**61**:899-910

[80] Mohandas N, Evans E. Mechanical properties of the red cell membrane in relation to molecular structure and genetic defects. Annual Review of Biophysics and Biomolecular Structure. 1994;**23**:787-818. DOI: 10.1146/annurev.bb.23.060194.004035

[81] Delaunay J. The molecular basis of hereditary red cell membrane disorders. Blood Reviews. 2007;**21**:1-20. DOI: 10.1016/j.blre.2006.03.005 Epub 2006 May 30

[82] Andolfo I, Russo R, Gambale A, Iolascon A. New insights on hereditary erythrocyte membrane defects. Haematologica. 2016;**101**:1284-1294. DOI: 10.3324/haematol.2016.142463 Epub 2016 Oct 18

[83] Perrotta S, Gallagher PG, Mohandas N. Hereditary spherocytosis. Lancet. 2008;**372**:1411-1426. DOI: 10.1016/S0140-6736(08)61588-3

[84] Safeukui I, Buffet PA, Deplaine G, Perrot S, Brousse V, Ndour A, et al. Quantitative assessment of sensing and sequestration of spherocytic erythrocytes by the human spleen. Blood. 2012;**120**:424-430. DOI: 10.1182/blood-2012-01-404103

[85] Clark MR, Mohandas N, Caggiano V, Shohet SB. Effects of abnormal cation transport on deformability of desiccytes. Journal of Supramolecular Structure.

1978;**8**:521-532. DOI: 10.1002/jss.400080414

[86] Barneaud-Rocca D, Pellissier B, Borgese F, Guizouarn H. Band 3 missense mutations and stomatocytosis: Insight into the molecular mechanism responsible for monovalent cation leak. International Journal of Cell Biology. 2011;**2011**:136802. DOI: 10.1155/2011/136802 Epub 2011 Aug 23

[87] Bogdanova A, Goede JS, Weiss E, Bogdanov N, Bennekou P, Bernhardt I, et al. Cryohydrocytosis: Increased activity of cation carriers in red cells from a patient with a band 3 mutation. Haematologica. 2010;**95**:189-198. DOI: 10.3324/haematol.2009.010215 Epub 2009 Dec 16

[88] Vives Corrons JL, Besson I, Aymerich M, Ayala S, Alloisio N, Delaunay J, et al. Hereditary xerocytosis: A report of six unrelated Spanish families with leaky red cell syndrome and increased heat stability of the erythrocyte membrane. British Journal of Haematology. 1995;**90**:817-822. DOI: 10.1111/j.1365-2141.1995.tb05201.x

[89] Fermo E, Vercellati C, Marcello AP, Zaninoni A, van Wijk R, Mirra N. Hereditary xerocytosis due to mutations in *PIEZO1* gene associated with heterozygous pyruvate kinase deficiency and beta-thalassemia trait in two unrelated families. Case Reports in Hematology. 2017;**2017**: 2769570. DOI: 10.1155/2017/2769570 Epub 2017 Mar 6

[90] Bae C, Gnanasambandam R, Nicolai C, Sachs F, Gottlieb PA. Xerocytosis is caused by mutations that alter the kinetics of the mechanosensitive channel PIEZO1. Proceedings of the National Academy of Sciences of the United States of America. 2013;**110**:E1162-E1168. DOI: 10.1073/pnas.1219777110 Epub 2013 Mar 4

[91] Archer NM, Shmukler BE, Andolfo I, Vandorpe DH, Gnanasambandam R, Higgins JM, et al. Hereditary xerocytosis revisited. American Journal of Hematology. 2014;**89**:1142-1146. DOI: 10.1002/ajh.23799 Epub 2014 Jul 21

[92] Longo D, Fauci A, Kasper D, Hauser S, Jameson J, Loscalzo J. Harrison's Principles of Internal Medicine. 18th ed. McGraw-Hill; 2013. Available from: www.accessmedicine.com

[93] van Solinge WW, van Wijk R. Erythrocyte enzyme disorders. In: Kaushansky K, Lichtman MA, Prchal JT, Levi MM, Press OW, Burns LJ, et al. Williams Haematology, 9th ed. McGraw-Hill; 2021

[94] Fischer DJ, Torrence NJ, Sprung RJ, Spence DM. Determination of erythrocyte deformability and its correlation to cellular ATP release using microbore tubing with diameters that approximate resistance vessels in vivo. Analyst. 2003;**128**:1163-1168. DOI: 10.1039/b308225n

[95] Weed RI, LaCelle PL, Merrill EW. Metabolic dependence of red cell deformability. The Journal of Clinical Investigation. 1969;**48**:795-809. DOI: 10.1172/JCI106038

[96] Makhro A, Huisjes R, Verhagen LP, Mañú-Pereira M, Llaudet-Planas E, Petkova-Kirova P, et al. Red cell properties after different modes of blood transportation. Frontiers in Physiology. 2016;**7**:288-293. DOI: 10.3389/fphys.2016.0028894

[97] Leblond PF, Coulombe L, Lyonnais J. Erythrocyte populations in pyruvate kinase deficiency anaemia following splenectomy. II. Cell deformability. British Journal of Haematology. 1978;**39**:

63-70. DOI: 10.1111/j.1365-2141.1978.tb07128.x

[98] Zaninoni A, Fermo E, Vercellati C, Consonni D, Marcello AP, Zanella A, et al. Use of laser assisted optical rotational cell analyzer (LoRRca MaxSis) in the diagnosis of RBC membrane disorders, enzyme defects, and congenital dyserythropoietic anemias: a monocentric study on 202 patients. Frontiers in Physiology. 2018;**9**:451-456. DOI: 10.3389/fphys.2018.00451

[99] Fermo E, Vercellati C, Marcello AP, Zaninoni A, Aytac S, Cetin M, et al. Clinical and molecular spectrum of glucose-6-phosphate isomerase deficiency. Report of 12 new cases. Frontiers in Physiology. 2019;**10**: 467-471. DOI: 10.3389/fphys.2019.00467

[100] Vives-Corrons JL, Krishnevskaya E, Rodriguez IH, Ancochea A. Characterization of hereditary red blood cell membranopathies using combined targeted next-generation sequencing and osmotic gradient ektacytometry. International Journal of Hematology. 2021;**113**:163-174. DOI: 10.1007/s12185-020-03010-9 Epub 2020 Oct 19

[101] Bruce LJ. Hereditary stomatocytosis and cation-leaky red cells–recent developments. Blood Cells, Molecules & Diseases. 2009;**42**:216-222. DOI: 10.1016/j.bcmd.2009.01.014 Epub 2009 Mar 3

[102] Fu M, Li L, Albrecht T, Johnson JD, Kojic LD, Nabi IR. Autocrine motility factor/phosphoglucose isomerase regulates ER stress and cell death through control of ER calcium release. Cell Death and Differentiation. 2011;**18**: 1057-1070. DOI: 10.1038/cdd.2010.181 Epub 2011 Jan 21